Lecture Notes in Mathematics

T0171975

C.I.M.E. Foundation Subseries

Volume 2263

More information about this subseries at http://www.springer.com/series/3114

Fondazione C.I.M.E., Firenze

C.I.M.E. stands for *Centro Internazionale Matematico Estivo*, that is, International Mathematical Summer Centre. Conceived in the early fifties, it was born in 1954 in Florence, Italy, and welcomed by the world mathematical community: it continues successfully, year for year, to this day.

Many mathematicians from all over the world have been involved in a way or another in C.I.M.E.'s activities over the years. The main purpose and mode of functioning of the Centre may be summarised as follows: every year, during the summer, sessions on different themes from pure and applied mathematics are offered by application to mathematicians from all countries. A Session is generally based on three or four main courses given by specialists of international renown, plus a certain number of seminars, and is held in an attractive rural location in Italy.

The aim of a C.I.M.E. session is to bring to the attention of younger researchers the origins, development, and perspectives of some very active branch of mathematical research. The topics of the courses are generally of international resonance. The full immersion atmosphere of the courses and the daily exchange among participants are thus an initiation to international collaboration in mathematical research.

C.I.M.E. Director (2002 – 2014)
Pietro Zecca
Dipartimento di Energetica "S. Stecco"
Università di Firenze
Via S. Marta, 3
50139 Florence
Italy
e-mail: zecca@unifi.it

C.I.M.E. Director (2015 –)
Elvira Mascolo
Dipartimento di Matematica "U. Dini"
Università di Firenze
viale G.B. Morgagni 67/A
50134 Florence
Italy
e-mail: mascolo@math.unifi.it

C.I.M.E. Secretary
Paolo Salani
Dipartimento di Matematica "U. Dini"
Università di Firenze
viale G.B. Morgagni 67/A
50134 Florence
Italy
e-mail: salani@math.unifi.it

CIME activity is carried out with the collaboration and financial support of INdAM (Istituto Nazionale di Alta Matematica)

For more information see CIME's homepage: **http://www.cime.unifi.it**

Ailana Fraser • André Neves • Peter M. Topping •
Paul C. Yang

Geometric Analysis

Cetraro, Italy 2018

Matthew J. Gursky • Andrea Malchiodi

Editors

FONDAZIONE
CIME
ROBERTO CONTI
CENTRO INTERNAZIONALE MATEMATICO ESTIVO
INTERNATIONAL MATHEMATICAL SUMMER CENTER

Authors

Ailana Fraser (iD)
Department of Mathematics
University of British Columbia
Vancouver, BC, Canada

Peter M. Topping (iD)
Department of Mathematics
University of Warwick
Coventry, UK

André Neves
Department of Mathematics
University of Chicago
Chicago, IL, USA

Paul C. Yang
Department of Mathematics
Princeton University
Princeton, NJ, USA

Editors

Matthew J. Gursky (iD)
Department of Mathematics
University of Notre Dame
Notre Dame, IN, USA

Andrea Malchiodi (iD)
Scuola Normale Superiore
Pisa, Italy

ISSN 0075-8434   ISSN 1617-9692   (electronic)
Lecture Notes in Mathematics
C.I.M.E. Foundation Subseries
ISBN 978-3-030-53724-1   ISBN 978-3-030-53725-8   (eBook)
https://doi.org/10.1007/978-3-030-53725-8

Mathematics Subject Classification: 53-XX, 35-XX

This Springer imprint is published by the registered company Springer Nature Switzerland AG.
The registered company address is: Gewerbestrasse 11, 6330 Cham, Switzerland

Preface

This volume contains the notes of the lectures delivered at the CIME Summer School *Geometric Analysis*, held during the week of June 18–22, 2018 in Cetraro (Cosenza). The school consisted of four courses: *Minimal Surfaces and Extremal Eigenvalue Problems on Surfaces* taught by Ailana Fraser (University of British Columbia), *Min–Max Theory, Volume Spectrum, and Minimal Surfaces* taught by Andre Neves (University of Chicago), *Ricci Flow and Ricci Limit Spaces* taught by Peter Topping (University of Warwick), and *CR Geometry in Three Dimensions* taught by Paul Yang (Princeton University). The courses covered a broad range of topics in geometric analysis.

The interaction of geometry and analysis has been incredibly fruitful. The resolution of the Geometrization Conjecture and the applications of gauge theory to four-dimensional topology are just two examples of remarkable advances that heavily relied on ideas and techniques from the analysis. In the other direction, conformal geometry and the theory of minimal surfaces have been the driving force behind many of the developments in the existence and regularity theory for nonlinear PDEs. The goal of the *Geometric Analysis* Summer School was to invite researchers of international stature to give lectures on areas that highlight this beautiful interaction. As with all CIME meetings, students and junior researchers benefited from attending the lectures and the mathematical discussions that they fostered.

The lectures of Ailana Fraser were on extremal eigenvalue problems and their connection to minimal surfaces. It is well known that the spectrum of the Laplacian contains a great deal of geometric information. A natural variational problem is to seek a metric on a fixed Riemannian manifold that maximizes the first eigenvalue over all metrics normalized to have the same volume. There is a remarkable connection between metrics that extremize the first eigenvalue and the existence of a minimal immersion of the manifold. Fraser's notes begin with an overview of some basic material on minimal surfaces in the sphere and free boundary minimal surfaces (FBMSs) in the unit ball. After reviewing classical eigenvalue estimates for both closed manifolds and for the Dirichlet and Neumann problems on manifolds with boundary, the Steklov eigenvalue problem is introduced. These are the eigenvalues of the Dirichlet-to-Neumann map and are one of the main topics of the lecture notes. Two main questions are posed: Given a smooth surface with boundary, does an extremal metric for the first Steklov eigenvalue exist? And if so, what are its geometric properties? The answer to the second question is that the eigenfunctions corresponding to an extremal metric on a surface M with boundary define an immersion that realizes M as an FBMS in the unit ball. After sketching a proof of this result, Fraser gives an outline of her existence work with R. Schoen, showing that any surface of genus zero and k boundary components admits an extremal metric. In the remainder of the notes, she discusses the problem in higher dimensions, as well as index bounds and the question of uniqueness. Fraser's notes demonstrate the beautiful interplay between spectral geometry and minimal surfaces.

The lectures of Andre Neves approach minimal surfaces from a different perspective and are coauthored by Fernando Marques (his collaborator on much of the work described in the notes). The Neves–Marques notes give an overview of recent spectacular advances in the existence theory for minimal surfaces via the minimax method, many of them motivated by Yau's conjecture that every closed Riemannian three-manifold contains infinitely many smooth, closed minimal surfaces. The problem has a long history, with the first general existence results obtained in papers by Almgren–Pitts and Schoen–Simon. Neves–Marques were able to settle Yau's conjecture for metrics of positive Ricci curvature, and more recently with Irie, they proved it for generic metrics, together with a density result. With Song, they were also able to prove equidistribution of minimal surfaces for generic metrics; combining a Morse-index bound from the authors, the proof of the *Multiplicity One conjecture* by Zhou, and a *Weyl law* for the volume spectrum proved by Liokumovich, they were able to prove existence of embedded, two-sided, minimal hypersurfaces with multiplicity one and given asymptotic volume. After recalling some basic concepts in Geometric Measure Theory, the authors describe the main tool used to prove their existence results, i.e., minimax theory. They then introduce the concept of *volume spectrum* (due to Gromow) of a Riemannian manifold and its Weyl's law, inspired by the asymptotic growth of eigenvalues, finally turning to prove the density and equidistribution theorems.

The notes of Peter Topping are on another important area of geometric analysis, the Ricci flow and its applications. Topping's goal is to use the Ricci flow to study non-compact manifolds with possibly unbounded curvature. In particular, one goal is to be able to run the Ricci flow for manifolds that arise as limits of sequences of metrics under certain curvature bounds (a *Ricci limit space*). The notes provide a very accessible introduction to the Ricci flow, using the two-dimensional case and the author's work with Giesen as a preparation for the higher dimensional setting. Topping starts by recalling some basic results in the closed case and then discussing other ones for complete manifolds with bounded curvature. After this, the situation for general initial metrics is considered, for which one may confront several difficulties, e.g., nonuniqueness of the flow, lack of completeness, or failure of maximum principles. Several examples are provided to illustrate each of these pathological occurrences. Despite these, he shows existence and uniqueness of the flow for arbitrary initial metrics within the class of complete flows, obtaining compactness results from some uniform local estimates for approximating flows. One beautiful continuation of this study is the proof (in joint works with Simon and McLeod) that in three dimensions, a Ricci limit space (which *a priori* is just a metric space) is actually bi-Hölder homeomorphic to a smooth manifold. This result is also proved by studying proper approximating Ricci flows and it is connected to a general conjecture by Anderson, Cheeger, Colding, and Tian.

The final contribution to this volume are the notes of Paul Yang on CR geometry. Yang and his collaborators have pioneered the approach of applying ideas and techniques from diverse fields such as conformal geometry, complex analysis, PDE techniques, and surface theory to prove important results on embeddability, the CR-Yamabe problem, and the study of minimal surfaces in sub-Riemannian geometry. Yang's notes are another excellent example of the interplay between geometry and analysis. After introducing some basic notions in CR geometry, Yang describes some relevant conformally covariant operators such as the conformal Laplacian and the Paneitz. He then introduces the concept of pseudo-Einstein contact form and its relation to the Szegö kernel, as shown by some fundamental work of Fefferman. He then turns to the embedding problem for general CR manifolds, describing the deep role of the CR Paneitz operator in the matter, proven in papers with Case, Chiu, and Chanillo. Surprisingly, the positivity of the operator and the embedding condition has also some role in the study of the CR-Yamabe problem and in the proof of a positive mass theorem in this setting: this is treated in works with Cheng and one of the Editors, where some relevant counterexamples are also given. Another unexpected phenomenon is the role of the pseudo-Einstein condition and the so-called Q'-*curvature* in the study of isoperimetric problems, shown in works with Wang. Such problems also motivate the study of surface theory in the CR setting, where the prescribed curvature problem is particularly challenging due to a lack of ellipticity: still, some interesting results are proved for the H-surface equation and counterparts for the Willmore energy in joint papers with Cheng, Hwang, and Zhang.

The quality of the lectures and the engagement of the students made for a lively and stimulating meeting. We are very grateful to the CIME Scientific Board for having approved our proposal and to the CIME staff for the valuable support during the organization and the running of the event. We are of course extremely grateful to the lecturers for the careful preparation of the courses and for their captivating lectures. Finally, we would also like to thank the participants for their interest in the course and their active involvement.

Notre Dame, IN, USA   Matthew J. Gursky
Pisa, Italy   Andrea Malchiodi

Contents

Chapter 1
Extremal Eigenvalue Problems and Free Boundary Minimal Surfaces in the Ball

Ailana Fraser

Abstract The main theme of this chapter is the study of extremal eigenvalue problems and its relations to minimal surface theory. We describe joint work with R. Schoen on progress that has been made on the Steklov eigenvalue problem for surfaces with boundary, and in higher dimensions. For surfaces, the Steklov eigenvalue problem has a close connection to free boundary minimal surfaces in Euclidean balls. Specifically, metrics that maximize Steklov eigenvalues are characterized as induced metrics on free boundary minimal surfaces in $\mathbb{B}^n$. We discuss the existence of maximizing metrics for surfaces of genus zero, and explicit characterizations of maximizing metrics for the annulus and Möbius band. We also give an overview of results on existence, uniqueness, and Morse index of free boundary minimal surfaces in $\mathbb{B}^n$.

1.1 Introduction

When we choose a metric on a manifold we determine the spectrum of the Laplace operator. Thus an eigenvalue may be considered as a functional on the space of metrics. For example the first eigenvalue would be the fundamental vibrational frequency. In some cases the normalized eigenvalues are bounded independent of the metric. In such cases it makes sense to attempt to find critical points in the space of metrics. For surfaces, the critical metrics turn out to be the induced metrics on certain special classes of minimal (mean curvature zero) surfaces in spheres and Euclidean balls. The eigenvalue extremal problem is thus related to other questions arising in the theory of minimal surfaces.

The focus of these lecture notes is on an extremal eigenvalue problem on manifolds with boundary. Given a compact Riemannian manifold with boundary, the most standard eigenvalue problems are the Dirichlet and Neumann eigenvalue

A. Fraser (✉)
Department of Mathematics, University of British Columbia, Vancouver, BC, Canada
e-mail: afraser@math.ubc.ca

M. J. Gursky, A. Malchiodi (eds.), *Geometric Analysis*, Lecture Notes
in Mathematics 2263, https://doi.org/10.1007/978-3-030-53725-8_1

problems. However, there is another important eigenvalue problem, and one that turns out to lead to an interesting variational problem geometrically, and that is the Steklov eigenvalue problem. We refer the reader to the recent article of A. Girouard and I. Polterovich [24] for a survey on the geometry of the Steklov problem. In this article we will describe joint work with R. Schoen on some progress that has been made on the Steklov eigenvalue problem for surfaces with boundary, and in higher dimensions. The extremals of this eigenvalue problem for surfaces are free boundary minimal surfaces in Euclidean balls, and we will give an overview of results on existence, uniqueness and Morse index of free boundary minimal surfaces in $\mathbb{B}^n$.

1.2 Minimal Submanifolds in $\mathbb{S}^n$ and Free Boundary Minimal Surfaces in $\mathbb{B}^n$

It is well known that minimal submanifolds of the sphere $\mathbb{S}^{n-1} \subset \mathbb{R}^n$ are characterized by the condition that the coordinate functions $x_1, \ldots, x_n$ are eigenfunctions of the Laplacian on the minimal submanifold, with eigenvalue equal to the dimension of the submanifold. Free boundary minimal submanifolds of the Euclidean unit ball $\mathbb{B}^n$ are characterized by the condition that the coordinate functions are Steklov eigenfunctions with eigenvalue 1. As a motivation for our study of the Steklov problem, we will start by reviewing these minimal submanifold characterizations.

Let Σ^k be a k-dimensional immersed submanifold in $\mathbb{R}^n$ with the induced metric. We denote by ∇ the connection on $\mathbb{R}^n$ and by ∇^Σ the induced connection on Σ. Given vector fields X, Y tangent to Σ, we have $\nabla_X Y = (\nabla_X Y)^\top + (\nabla_X Y)^\perp = \nabla_X^\Sigma Y + A(X, Y)$, where $(\cdot)^\top$ and $(\cdot)^\perp$ denote the components tangent and normal to Σ respectively, and A is the second fundamental form of Σ in $\mathbb{R}^n$. Let $e_1, \ldots, e_k$ be a local orthonormal frame tangent to Σ. If $x = (x_1, \ldots, x_n)$ is the position vector, observe that for any $v \in \mathbb{R}^n$, $\nabla_v x = v$ and

$$\Delta_\Sigma x = \sum_{i=1}^{k} \left(\nabla_{e_i} \nabla_{e_i} x - \nabla_{\nabla_{e_i}^\Sigma e_i} x \right) = \sum_{i=1}^{n} \nabla_{e_i} e_i - \nabla_{e_i}^\Sigma e_i = \sum_{i=1}^{k} A(e_i, e_i) = H$$

where H is the mean curvature of Σ in $\mathbb{R}^n$. A minimal submanifold is a submanifold with zero mean curvature. Thus,

$$\Sigma^k \subset \mathbb{R}^n \text{ is minimal} \quad \Longleftrightarrow \quad H \equiv 0 \quad \Longleftrightarrow \quad \Delta_\Sigma x_i = 0, \quad i = 1, \ldots, n. \tag{1.1}$$

That is, minimal submanifolds of $\mathbb{R}^n$ are characterized by the condition that the coordinate functions $x_1, \ldots, x_n$ are harmonic functions on Σ.

Now suppose that Σ^k is a k-dimensional immersed submanifold in $\mathbb{S}^n \subset \mathbb{R}^{n+1}$, with the induced metric. Let $e_1, \ldots, e_{n+1}$ be an adapted local orthonormal frame, such that $e_1, \ldots, e_k$ are tangent to Σ, $e_{k+1} = x$ is the outward unit normal to $\mathbb{S}^n$, and $e_{k+2}, \ldots, e_{n+1}$ are normal to Σ and tangent to $\mathbb{S}^n$. We let $\nabla^{\mathbb{S}^n}$ denote the induced connection on $\mathbb{S}^n$, $H_{\Sigma \subset \mathbb{R}^{n+1}}$ denote the mean curvature of Σ in $\mathbb{R}^{n+1}$ and $H_{\Sigma \subset \mathbb{S}^n}$ denote the mean curvature of Σ in $\mathbb{S}^n$. Then,

$$H_{\Sigma \subset \mathbb{R}^{n+1}} = \sum_{i=1}^{k} \left(\nabla_{e_i} e_i \right)^{\perp} = \sum_{i=1}^{k} \left[\langle \nabla_{e_i} e_i, x \rangle x + \sum_{\alpha=k+2}^{n+1} \langle \nabla_{e_i} e_i, e_\alpha \rangle e_\alpha \right].$$

Notice that $\langle \nabla_{e_i} e_i, x \rangle = -1$ since it is the second fundamental form of $\mathbb{S}^n$ in $\mathbb{R}^{n+1}$, with respect to the outward unit normal, in the direction of the unit vector e_i. Using this, and since e_α, $\alpha = k + 2, \ldots, n + 1$, form an orthonormal basis normal to Σ and tangent to $\mathbb{S}^n$, we have

$$H_{\Sigma \subset \mathbb{R}^{n+1}} = -kx + \sum_{i=1}^{k} \sum_{\alpha=k+2}^{n+1} \langle \nabla_{e_i}^{\mathbb{S}^n} e_i, e_\alpha \rangle e_\alpha = -kx + H_{\Sigma \subset \mathbb{S}^n}.$$

Therefore,

$$\Sigma^k \text{ is minimal in } \mathbb{S}^n \iff H_{\Sigma \subset \mathbb{S}^n} \equiv 0$$
$$\iff H_{\Sigma \subset \mathbb{R}^{n+1}} = -kx$$
$$\iff \Delta_\Sigma x_i + k x_i = 0, \quad i = 1, \ldots, n + 1.$$

That is, k-dimensional minimal submanifolds of the sphere $\mathbb{S}^n$ are characterized by the condition that the coordinate functions $x_1, \ldots, x_{n+1}$ are eigenfunctions of the Laplacian on Σ, with eigenvalue k.

If $C(\Sigma)$ is the cone over $\Sigma \subset \mathbb{S}^n \subset \mathbb{R}^{n+1}$,

$$C(\Sigma) := \{ x \in \mathbb{R}^{n+1}, x \neq 0 : \frac{x}{|x|} \in \Sigma \} \cup \{0\}$$

then we leave it to the reader to check that Σ is minimal in $\mathbb{S}^n$ if and only if $C(\Sigma)$ is minimal in $\mathbb{R}^{n+1}$. Note that $C(\Sigma)$ meets $\mathbb{S}^n$ orthogonally. Thus, if Σ is minimal in $\mathbb{S}^n$, then $C(\Sigma) \cap \overline{\mathbb{B}}^{n+1}$ is a free boundary minimal submanifold in $\mathbb{B}^{n+1}$, possibly with a singularity at the origin.

In general, suppose Σ^k is a k-dimensional immersed submanifold in $\mathbb{B}^n$ with nonempty boundary contained in the boundary of the ball $\partial\Sigma \subset \partial\mathbb{B}^n$, with the induced metric. We say that Σ is a *free boundary minimal submanifold of* $\mathbb{B}^n$ if Σ is minimal ($H = 0$) and Σ meets the boundary of the ball orthogonally. Free boundary minimal submanifolds of $\mathbb{B}^n$ arise variationally as critical points of the volume functional among submanifolds of the ball with boundary on the boundary of the ball (see Sect. 1.5.3). By (1.1), the condition that $\Sigma \subset \mathbb{B}^n$ is minimal is equivalent to the condition that the coordinate functions are harmonic functions on Σ. The condition that Σ meets the boundary of the ball orthogonally is equivalent to the condition that the outward unit conormal η of Σ along $\partial\Sigma$ is orthogonal to the sphere, or $\eta = x$. That is,

$$\nabla_\eta x = \eta = x.$$

Hence, $\Sigma \subset \mathbb{B}^n$ is minimal and meets $\partial \mathbb{B}^n$ orthogonally, if and only if

$$\begin{cases} \Delta_\Sigma x_i = 0 & \text{on } \Sigma \\ \frac{\partial x_i}{\partial \eta} = x_i & \text{on } \partial \Sigma \end{cases}$$

for $i = 1, \ldots, n$. That is, free boundary minimal submanifolds of $\mathbb{B}^n$ are characterized by the condition that the coordinate functions are Steklov eigenfunctions (see Sect. 1.4) with eigenvalue 1.

We will see in Sects. 1.3 and 1.4 that the connection between these classes of minimal surfaces and the corresponding eigenvalue problems is in fact much deeper than we have observed here.

1.3   Preliminaries on Eigenvalue Bounds

For compact surfaces without boundary the Laplace eigenvalue problem leads to an interesting extremal eigenvalue problem, for which maximizing metrics are known to exist in certain cases, and for which extremal metrics have an interesting geometric characterization as induced metrics on minimal surfaces in spheres. We outline this in Sect. 1.3.1 below. For compact Riemannian manifolds with boundary, the most standard eigenvalue problems are the Dirichlet and Neumann eigenvalue problems. However, as we will discuss in Sects. 1.3.2 and 1.3.3, these eigenvalue problems do not directly lead in the same way to interesting extremal eigenvalue problems.

1.3.1   Compact Manifolds Without Boundary

Given a compact manifold M without boundary, the choice of a Riemannian metric g on M gives a Laplace operator Δ_g, which has a discrete set of eigenvalues $\lambda_0 = 0 < \lambda_1 \le \lambda_2 \le \ldots$. One can then ask how the eigenvalues are related to the geometry of (M, g). There are many known eigenvalue bounds with most requiring geometric assumptions such as curvature, volume, or diameter bounds. In a few cases there are bounds depending only on the topology of M which are metric independent. The following result was proven by J. Hersch in 1970.

Theorem 1.3.1 ([25]) *For any smooth metric g on $\mathbb{S}^2$*

$$\lambda_1(g) A_g(\mathbb{S}^2) \le 8\pi$$

with equality if and only if g is a constant curvature metric.

Note that for the unit sphere the area is 4π and $\lambda_1 = 2$ with the coordinate functions being first eigenfunctions. In general, for surfaces, there is an upper bound that depends only on the genus of the surface. In 1980 Yang-Yau proved:

Theorem 1.3.2 ([54]) *For any smooth metric g on the closed orientable surface M of genus* γ

$$\lambda_1(g)A_g(M) \leq 8\pi \left[\frac{\gamma+3}{2}\right].$$

For non-orientable surfaces, in 2016 Karpukhin proved:

Theorem 1.3.3 ([30]) *For any smooth metric g on the closed nonorientable surface M of genus* γ

$$\lambda_1(g)A_g(M) < 16\pi \left[\frac{\gamma+3}{2}\right].$$

The genus of a nonorientable surface is defined to be the genus of its orientable double cover. Recently Karpukhin [32] has shown that the Yang-Yau inequality of Theorem 1.3.2 is strict for all $\gamma > 2$, and in general, Theorems 1.3.2 and 1.3.3 are not expected to be sharp upper bounds. The explicit value of the sharp upper bound has only been computed for a few surfaces. For $\mathbb{S}^2$ the value is 8π and the constant curvature metric is the unique maximum by Theorem 1.3.1. For $\mathbb{RP}^2$ the value is 12π and the constant curvature metric is the unique maximum by a result of Li and Yau [37] from the 1982. They made a connection with minimal surfaces in spheres; in particular, the Veronese minimal embedding of $\mathbb{RP}^2$ into $\mathbb{S}^4$ plays a key role in the proof. For $\mathbb{T}^2$ the value is $8\pi/\sqrt{3}$ and the flat metric on the 60^0 rhombic torus is the unique maximum by a result of Nadirashvili [41] from 1996. This surface can be isometrically minimally embedded into $\mathbb{S}^5$ by first eigenfunctions. For the Klein bottle the extremal metric is smooth and unique but not flat. This follows from work of Nadirashvili [41] from 1996 on existence of a maximizer, Jakobson et al. [27] from 2006 who constructed the metric, and El Soufi et al. [14] from 2006 who proved it is unique. The metric arises on a minimal immersion of the Klein bottle into $\mathbb{S}^4$. For the surface of genus 2 the value is 16π and the maximal metric is a singular metric of constant positive curvature with 6 singular points, that arises from a specific complex structure and a branched double cover of $\mathbb{S}^2$ [26, 42].

In general, Nadirashvili established a connection between maximizing metrics on closed surfaces and minimal surfaces in spheres.

Theorem 1.3.4 ([41]) *Let M be a compact surface without boundary, and suppose* g_0 *is a metric on M such that*

$$\lambda_1(g_0)A_{g_0}(M) = \sup_g \lambda_1(g)A_g(M)$$

where the supremum is over all smooth metrics on M. Then there exist independent first eigenfunctions $u_1, \ldots, u_{n+1}$, $n \geq 2$, which, after rescaling the metric, give an isometric minimal immersion $u = (u_1, \ldots, u_{n+1})$ of M into the unit sphere $\mathbb{S}^n$.

1.3.2　Manifolds with Boundary: The Dirichlet Problem

Given a compact Riemannian manifold (M, g) with boundary $\partial M \neq 0$, the Dirichlet eigenvalue problem

$$\begin{cases} \Delta_g u = 0 & \text{on } M \\ u = 0 & \text{on } \partial M \end{cases}$$

has a discrete set of eigenvalues $0 < \lambda_1 < \lambda_2 \leq \lambda_3 \leq \cdots \to \infty$. The first eigenvalue λ_1 is characterized variationally by

$$\lambda_1 = \inf \left\{ \frac{\int_M |\nabla u|^2 \, d\mu_g}{\int_M u^2 \, d\mu_g} : u \in W_0^{1,2}(M), \, u \neq 0 \right\}.$$

If we consider domains in $\mathbb{R}^n$, there is a sharp lower bound on the first Dirichlet eigenvalue λ_1 depending only on the volume of the domain. Specifically, the Faber-Krahn inequality asserts that a ball is the unique minimizer of λ_1 among all domains with the same volume:

Theorem 1.3.5 (Faber-Krahn Inequality) *For any bounded domain Ω in $\mathbb{R}^n$,*

$$\lambda_1(\Omega) \geq \lambda_1(\Omega^*)$$

where Ω^* is a ball with the same volume as Ω. Equality holds if and only if Ω is a ball.

It is easy to see that there is no upper bound on λ_1 in terms of volume alone.

If we allow geometries that are not domains, then even for the two dimensional disk we see that there is no positive lower bound on λ_1 in terms of the area. For example, as can be seen from the variational characterization of λ_1, the complement of a small disk in the two sphere has λ_1 which is arbitrarily small and goes to zero as the radius of the disk goes to zero.

1.3.3 Manifolds with Boundary: The Neumann Problem

Given a compact Riemannian manifold (M, g) with boundary $\partial M \neq 0$, the Neumann eigenvalue problem

$$\begin{cases} \Delta_g u = 0 & \text{on } M \\ \frac{\partial u}{\partial \eta} = 0 & \text{on } \partial M \end{cases}$$

has a discrete set of eigenvalues $0 = \mu_0 < \mu_1 \leq \mu_2 \leq \cdots \to \infty$. The first nonzero eigenvalue μ_1 is characterized variationally by

$$\mu_1 = \inf \left\{ \frac{\int_M |\nabla u|^2 \, d\mu_g}{\int_M u^2 \, d\mu_g} : u \in W^{1,2}(M), \int_M u \, d\mu_g = 0 \right\}.$$

For domains in $\mathbb{R}^n$ there is a sharp upper bound on the first Neumann eigenvalue μ_1 in terms of the volume. It was shown by G. Szegö in 1954 that the unit disk maximizes μ_1 among simply connected plane domains with area π.

Theorem 1.3.6 ([49]) *For any simply-connected bounded domain $\Omega \subset \mathbb{R}^2$*

$$\mu_1(\Omega)A(\Omega) \leq \mu_1(\mathbb{D})A(\mathbb{D})$$

with equality if and only if Ω is a disk.

This result was generalized in 1956 by H. Weinberger to arbitrary domains in $\mathbb{R}^n$.

Theorem 1.3.7 ([52]) *For any bounded domain Ω in $\mathbb{R}^n$*

$$\mu_1(\Omega) \leq \mu_1(\Omega^*)$$

where Ω^* is a ball with the same volume as Ω. Equality holds if and only if Ω is a ball.

The same method used in Theorem 1.3.2 for closed surfaces implies an upper bound on $\mu_1(g)A_g(M)$ for any smooth metric g on a surface M of genus γ with boundary:

$$\mu_1(g)A_g(M) \leq 8\pi \left[\frac{\gamma + 3}{2} \right].$$

However in this case, in contrast to the situation for closed surfaces, it is unlikely that the problem will have maximizers. If we take our surface to be a disk we get the Hersch bound $\mu_1 A \leq 8\pi$ and the bound is asymptotically achieved by the two-sphere minus a very small ball.

There is a third important eigenvalue problem on manifolds with boundary, the Steklov problem, that has the correct properties, as we will discuss in the next section.

1.4   The Steklov Eigenvalue Problem

Let (M^n, g) be a compact n-dimensional Riemannian manifold with boundary. A function u on M is a *Steklov eigenfunction* with eigenvalue σ if

$$\begin{cases} \Delta_g u = 0 & \text{on } M \\ \frac{\partial u}{\partial \eta} = \sigma u & \text{on } \partial M. \end{cases}$$

Steklov eigenvalues are eigenvalues of the *Dirichlet-to-Neumann operator*. The Dirichlet-to-Neumann operator is an operator on functions on the boundary of the manifold

$$L : C^\infty(\partial M) \to C^\infty(\partial M)$$

given by

$$Lu = \frac{\partial \hat{u}}{\partial \eta}$$

where $\hat{u}$ is the harmonic extension of u to M:

$$\begin{cases} \Delta_g \hat{u} = 0 & \text{on } M \\ \hat{u} = u & \text{on } \partial M. \end{cases}$$

L is a self-adjoint operator which is non-negative definite, since by Green's formulas

$$\int_{\partial M} (u L(v) - v L(u)) \, d\mu_{\partial M} = \int_M [\hat{u} \Delta(\hat{v}) - \hat{v} \Delta(\hat{u})] \, d\mu_M = 0,$$

and

$$\int_{\partial M} u L(u) \, d\mu_{\partial M} = \int_M |\nabla \hat{u}|^2 \, d\mu_M.$$

Recall that the trace embedding from $W^{1,2}(M)$ to $L^2(\partial M)$ is compact. This makes it possible to diagonalize the quadratic form $\int_M |\nabla u|^2 \, d\mu_M$ on the unit sphere in $L^2(\partial M)$ to construct the eigenvalues

$$0 = \sigma_0 < \sigma_1 \leq \sigma_2 \leq \sigma_3 \leq \cdots \to \infty$$

and orthonormal eigenfunctions $u_0, u_1, u_2, u_3, \ldots$ satisfying

$$\Delta_g u_i = 0 \text{ on } M, \text{ and } \frac{\partial u_i}{\partial \eta} = \sigma_i u_i \text{ on } \partial M.$$

Note that $\sigma_0 = 0$ and u_0 is a constant function while $\sigma_1 > 0$ (we take M to be connected).

Thus the eigenfunctions are critical points of the Rayleigh quotient

$$R(u) = \frac{\int_M |\nabla u|^2 \, d\mu_M}{\int_{\partial M} u^2 \, d\mu_{\partial M}}$$

among $W^{1,2}$ functions on M. The j-th Steklov eigenvalue σ_j is characterized variationally by:

$$\sigma_j = \inf\{ R(u) : u \in W^{1,2}(M), \int_{\partial M} uu_i \, d\mu_{\partial M} = 0, \ i = 0, 1, \ldots, j-1\}.$$

The eigenfunction is then a function that achieves the infimum. By elliptic regularity the eigenfunctions are smooth on the closure of M. Alternatively,

$$\sigma_j = \min_{E \in \mathcal{E}(j)} \sup_{0 \neq u \in E} R(u) \tag{1.2}$$

where $\mathcal{E}(j)$ is the set of all j-dimensional subspaces of $W^{1,2}(M)$ that are L^2-orthogonal to the constants on ∂M.

For surfaces, the Steklov eigenvalues satisfy a natural conformal invariance.

Definition 1.4.1 We say that two metrics g_1 and g_2 on a surface M are σ-*homothetic* if there is a conformal diffeomorphism $F : (M, g_1) \to (M, g_2)$ that is a homothety on the boundary; that is, $F^*g_2 = \lambda^2 g_1$ with λ constant on ∂M.

Since the energy is conformally invariant for maps from surfaces, we see from the variational characterization of the Steklov eigenvalues that if g_1 and g_2 are σ-homothetic then they have the same normalized Steklov eigenvalues; that is, $\sigma_i(g_1)L_{g_1}(\partial M) = \sigma_i(g_2)L_{g_2}(\partial M)$ for each i.

Example 1.4.1 The Steklov eigenvalues of the Euclidean ball $\mathbb{B}^n$ are the nonnegative integers $0, 1, 2, \ldots$, and the eigenspace corresponding to eigenvalue k is the space of homogeneous harmonic polynomials of degree k. In particular, $\sigma_1 = 1$ and the eigenspace is spanned by the coordinate functions $x_1, \ldots, x_n$.

More generally if we take a metric cone $M = [0, 1] \times M_0$ for a compact $n-1$ manifold (M_0, g_0) with metric $g = dr^2 + r^2 g_0$ then the Steklov eigenfunctions are homogeneous extensions of the eigenfunctions of (M_0, g_0) with the unique homogeneity which makes them harmonic and bounded near $r = 0$. In this sense the Steklov problem includes the spectral problem for compact manifolds. The singularity at $r = 0$ is mild enough that the analysis works on M.

Example 1.4.2 For a thin rectangle $\Omega_\epsilon := (-1, 1) \times (-\epsilon, \epsilon)$, $0 < \epsilon \ll 1$, considering pairwise orthogonal functions $u_i(x, y) = \sin(\pi i x)$, $i \in \mathbb{N}$, we see using the variational characterization (1.2) that

$$\lim_{\epsilon \to 0} \sigma_j(\Omega_\epsilon) = 0.$$

More generally, any family M_ϵ of manifolds having a thin collapsing passage has $\sigma_j(\Omega_\epsilon)$ becoming arbitrarily small as $\epsilon \to 0$ ([23]). In particular, there is no lower bound in terms of volume alone.

The focus of this article is on sharp upper bounds on the first Steklov eigenvalue, and geometries that maximize the first eigenvalue. Specifically, we will be interested in thinking of the Steklov eigenvalues as functions of the metric g. Given a compact n-dimensional manifold M with boundary, the choice of a Riemannian metric g gives a Laplace operator Δ_g and associated Dirichlet-to-Neumann operator, which has a discrete set of eigenvalues

$$\sigma_0 = 0 < \sigma_1(g) \leq \sigma_2(g) \leq \cdots \leq \sigma_k(g) \leq \cdots$$

A basic question is, how big can the first eigenvalue be? That is, if we consider the functional that assigns to any smooth metric its first nonzero eigenvalue

$$g \mapsto \sigma_1(g)$$

then we can ask to find a metric that maximizes $\sigma_1(g)$. If we hope to be able maximize $\sigma_1(g)$, the first thing we would need to know is that there is some upper bound. Observe that the eigenvalues are not scale invariant. If we scale the metric by a positive constant $c > 0$, then

$$\sigma_1(cg) = \frac{1}{\sqrt{c}}\sigma_1(g).$$

Thus, some type of normalization is needed. Since the Dirichlet-to-Neumann operator is an operator on functions on the boundary, it is natural to normalize the boundary volume and restrict to metrics of fixed boundary volume $|\partial M|_g = 1$, or equivalently consider the normalized eigenvalues

$$\overline{\sigma}_j(g) := \sigma_j(g) \, |\partial M|_g^{\frac{1}{n-1}}$$

which are scale invariant.

1.4.1 Coarse Upper Bounds for Steklov Eigenvalues on Surfaces

There is a very classical estimate for Stekov eigenvalues that goes back to 1954, due to Weinstock.

Theorem 1.4.1 ([53]) *For any simply-connected bounded domain $\Omega \subset \mathbb{R}^2$*

$$\sigma_1(\Omega)L(\partial\Omega) \leq 2\pi = \sigma_1(\mathbb{D})L(\partial\mathbb{D})$$

with equality if and only if Ω is a disk.

Another way to say this is that the disk uniquely maximizes σ_1 among simply connected domains with the same boundary length. More generally the inequality holds for any simply connected surface with a metric. This is equivalent to allowing a positive density function ρ on $\partial\Omega$. The same proof gives the stronger bound

$$\frac{1}{\sigma_1} + \frac{1}{\sigma_2} \geq \frac{L}{\pi}.$$

We note that Theorem 1.4.1 is parallel to Szegö's estimate for the first Neumann eigenvalue (Theorem 1.3.6), and the proofs are similar, as we outline here.

Proof Since Ω is simply-connected, by the Riemann mapping theorem, there is a conformal diffeomorphism $\varphi : \Omega \to \mathbb{D}$. The idea of the proof is to use the component functions of φ as test functions in the variational characterization of σ_1. However in order to be a valid test function, a function must be L^2-orthogonal to the constant functions on the boundary $\partial\Omega$. It is well known that there is enough flexibility in the group of conformal diffeomorphisms of the disk that there exists a conformal diffeomorphism $F : \mathbb{D} \to \mathbb{D}$ such that

$$\int_{\partial\Omega} (F \circ \varphi)\, ds = 0.$$

Let $u = (u_1, u_2) = F \circ \varphi : \Omega \to \mathbb{D}$. Then $\int_{\partial\Omega} u_i\, ds = 0$ for $i = 1, 2$, and by the variational characterization of the first nonzero eigenvalue σ_1,

$$\sigma_1 \int_{\partial\Omega} u_i^2\, ds \leq \int_\Omega |\nabla u_i|^2\, da.$$

Summing over $i = 1, 2$ we have

$$\sigma_1 \int_{\partial\Omega} \sum_{i=1}^2 u_i^2\, ds \leq \int_\Omega \sum_{i=1}^2 |\nabla u_i|^2\, da$$

$$\sigma_1 \int_{\partial\Omega} ds \leq \int_\Omega |\nabla u|^2\, da$$

$$\sigma_1 L(\partial\Omega) \leq 2\,A(u(\Omega)) = 2A(\mathbb{D}) = 2\pi$$

where we have used the fact that for a conformal map, the energy of the map is twice the area of its image. □

It is straightforward to construct annuli that violate the Weinstock estimate, so the same bound does not hold for more general regions in the plane.

Example 1.4.3 Consider a Euclidean annulus $A_\epsilon := \mathbb{D}_1 \setminus \mathbb{D}_\epsilon$, $0 < \epsilon \ll 1$, where $\mathbb{D}_r$ denote the disk of radius r centered at the origin in $\mathbb{R}^2$. Using separation of variables, one can explicitly calculate the Steklov eigenvalues ([12], [24, Example 4.2.5]) and see that for $\epsilon > 0$ small enough

$$\sigma_1(A_\epsilon)L(\partial A_\epsilon) > \sigma_1(\mathbb{D})L(\partial \mathbb{D}).$$

Although Weinstock's inequality is no longer true for non-simply-connected planar domains, the supremum of the first normalized Steklov eigenvalue $\sigma_1 L$ among all planar domains is finite. In fact, an extension of the idea of the proof gives an upper bound for arbitrary Riemannian surfaces with boundary.

Theorem 1.4.2 ([17, 34]) *Let (M, g) be a compact orientable Riemannian surface of genus γ with $k > 0$ boundary components. Then,*

$$\sigma_1(g)\, L_g(\partial M) \;\leq\; \min\left\{2\pi(\gamma + k),\; 8\pi\left[\frac{\gamma + 3}{2}\right]\right\}.$$

Proof Outline The proof follows the approach of the proof of Theorem 1.4.1.

(i) The first bound uses the existence of a proper conformal branched covering $\varphi : M \to \mathbb{D}$ with $\deg(\varphi) \leq \gamma + k$.
(ii) For the second bound, fill the boundary components of M with disks to get a closed surface $\overline{M}$ of genus γ, and use existence of a conformal branched covering $\varphi : \overline{M} \to \mathbb{S}^2$ with $\deg(\varphi) \leq [(\gamma + 3)/2]$.

In each case we may then balance the map as before, and use the component functions as test functions in the variational characterization of σ_1 to obtain the upper bounds.   □

Remark 1.4.1 The first bound gives a sharper upper bound when the number of boundary components k is small, but as k becomes large enough the second bound, which depends only on the genus γ, gives a sharper upper bound. We note that there exist surfaces with arbitrarily large normalized first Steklov eigenvalue; that is, there is no upper bound independent of the genus. Specifically, it was shown in [10] that for each $\gamma \in \mathbb{N}$ there exists a compact surface M_γ of genus γ with connected boundary such that

$$\sigma_1(M_\gamma)L(\partial M_\gamma) \geq C(\gamma - 1)$$

for some constant $C > 0$.

When $\gamma = 0$ and $k = 1$, for M is simply-connected, Theorem 1.4.2 reduces to Weinstock's estimate and is sharp. However we expect these to be non-sharp bounds in general. In fact, it follows from [17, Theorem 2.5] that the first upper bound is not sharp for $\gamma = 0$ and $k > 1$. A basic question is, *what is the sharp upper bound for other surfaces?*

The fact that there is an upper bound on the first normalized eigenvalue $\sigma_1 L$ of any compact orientable Riemannian surface (M, g), that is independent of the metric and depends only on the topology of the surface, suggests the following extremal problem. Given a compact orientable surface M of genus γ with k boundary components, let

$$\sigma^*(\gamma, k) := \sup\{\sigma_1(g)L_g(\partial M) \ : \ g \text{ smooth metric on } M\}.$$

Theorem 1.4.1 implies that $\sigma^*(0, 1) = 2\pi$ and the maximizing metric is the flat metric on the unit disk. The question we would like to address is, what are the maximizing metrics for other surfaces with boundary? To be precise:

Basic Questions *Given a smooth surface M with boundary,*

1. *Does there exist a metric that maximizes $\sigma_1 L$?*
2. *If so what can we say about its geometry?*

We will start by considering the second question.

1.4.2   Characterization of Maximizing Metrics

The following result shows that a metric that maximizes $\sigma_1 L$ arises from a free boundary minimal surface in $\mathbb{B}^n$ for some $n \geq 2$.

Theorem 1.4.3 ([20]) *Let M be a compact surface with boundary, and suppose g_0 is a smooth metric on M such that*

$$\sigma_1(g_0)L_{g_0}(\partial M) = \sup_g \sigma_1(g)L_g(\partial M)$$

where the supremum is over all smooth metrics on M. Then the multiplicity of $\sigma_1(g_0)$ is at least 2 and, after rescaling g_0 so that $\sigma_1(g_0) = 1$, there exist first Steklov eigenfunctions $u_1, \ldots, u_n, n \geq 2$, such that

$$u = (u_1, \ldots, u_n) : M \to \mathbb{B}^n$$

is a proper conformal branched immersion that is an isometry on ∂M. In particular, $u(M)$ is minimal and meets $\partial \mathbb{B}^n$ orthogonally, and the maximizing metric g_0 is the induced metric on ∂M.

The result says that any maximizing metric is "σ-homothetic" (see Definition 1.4.1) to the induced metric from a free boundary minimal immersion of the surface M into the ball $\mathbb{B}^n$, for some $n \geq 2$, by first Steklov eigenfunctions. Recall from Sect. 1.2 that a *free boundary minimal surface in $\mathbb{B}^n$* is a minimal surface in

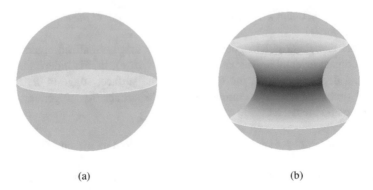

Fig. 1.1 (**a**) Equatorial disk. (**b**) Critical catenoid. Images by Emma Fajeau. First published in *Notices Amer. Math. Soc.* 65, no. 11 (2018), published by the American Mathematical Society

the ball $\mathbb{B}^n$ that meets the boundary of the ball $\partial \mathbb{B}^n$ orthogonally. Recall also that

$$\Sigma \subset \mathbb{R}^n \text{ minimal} \iff \Delta_\Sigma x_i = 0 \text{ for } i = 1, \ldots, n$$

$$\Sigma \text{ meets } \partial \mathbb{B}^n \text{ orthogonally} \iff \frac{\partial x_i}{\partial \eta} = x_i \text{ for } i = 1, \ldots, n$$

and so free boundary minimal surfaces in $\mathbb{B}^n$ are characterized by the condition that the coordinate functions $x_1, \ldots, x_n$ are Steklov eigenfunctions with eigenvalue one. Theorem 1.4.3 shows that the connection between free boundary minimal surfaces in a ball and the Steklov eigenvalue problem is much deeper than this, and in fact free boundary minimal surfaces in $\mathbb{B}^n$ are extremals for the Steklov eigenvalues.

Example 1.4.4 The simplest example of a free boundary minimal surface in a ball is an equatorial plane disk in $\mathbb{B}^3$ (Fig. 1.1(a)). It follows from Theorem 1.4.1 that the induced metric on an equatorial disk maximizes $\sigma_1(g)L_g(\partial M)$ among all smooth metrics g on a simply connected surface M.

Example 1.4.5 The *critical catenoid* is the unique piece of a suitable scaled catenoid in $\mathbb{R}^3$ that lies inside the unit ball $\mathbb{B}^3$ and meets the boundary of the ball orthogonally (Fig. 1.1(b)). Explicitly, the critical catenoid is given by the embedding

$$u : [-T_0, T_0] \times \mathbb{S}^1 \to \mathbb{B}^3$$

with

$$u(t, \theta) = \frac{1}{\sqrt{\cosh^2 T_0 + T_0^2}} (\cosh t \cos \theta, \cosh t \sin \theta, t)$$

where T_0 is the unique positive solution of $\coth t = t$. It is proved in [20], as will be discussed in Sect. 1.4.3, that the induced metric on the critical catenoid maximizes the first normalized Steklov eigenvalue among all smooth metrics on the annulus.

Theorem 1.4.3 should be compared with Theorem 1.3.4, and the proof involves some similar ideas. We will give an outline of the proof, and refer the reader to [20, Proposition 5.2] for full details, and to [18] for generalizations. The proof is like computing the Euler-Lagrange equation for the functional $g \mapsto \sigma_1(g)$ on the space of smooth metrics with fixed boundary length, except that $\sigma_1(g)$ is not a smooth function of the metric g, because of multiplicity. Although it is not smooth, the σ_1 functional is Lipschitz [20, Lemma 5.1].

Outline of Proof of Theorem 1.4.3 The proof has three main steps.

Step 1 Compute the first variation of the eigenvalue $\sigma_1(t)$ at points where the derivative exists, for a smooth path of metrics $g(t)$ with fixed boundary length. If we denote derivatives at $t = 0$ with 'dots' we let $h = \dot{g}$, and the length constraint on the boundary translates to

$$\int_{\partial M} h(T, T)\, ds = 0$$

where T denotes the oriented unit tangent vector to ∂M.

If the derivative of $\sigma_1(t)$ exists at $t = 0$, it may be computed most easily by differentiating at $t = 0$

$$F(t) = \int_M |\nabla u_t|_t^2\, da_t - \sigma_1(t) \int_{\partial M} u_t^2\, ds_t$$

where u_t is a smoothly varying path of functions such that u_0 is a first eigenfunction for $g(0)$ with $\int_{\partial M} u_0^2\, ds_0 = 1$ and $\int_{\partial M} u_t\, ds_t = 0$. By the variational characterization of $\sigma_1(t)$, we have $F(0) = 0$ and $F(t) \geq 0$, so $F'(0) = 0$.

Denoting the derivative of $\sigma_1(t)$ at $t = 0$ by $\dot{\sigma}_1$, we compute that

$$\dot{\sigma}_1 = -\int_M \langle \tau(u), h \rangle\, da - \int_{\partial M} u^2 h(T, T)\, ds$$

where $\tau(u)$ is the stress-energy tensor of u given by

$$\tau(u)_{ij} = \partial_i u \partial_j u - \frac{1}{2}|\nabla u|^2 g_{ij}.$$

Step 2 Assuming that $\sigma_1 L$ is maximized for g_0, show that for any variation h there exists an eigenfunction u (depending on h) with

$$Q_h(u) := \int_M \langle \tau(u), h \rangle\, da + \int_{\partial M} u^2 h(T, T)\, ds = 0.$$

This is accomplished by using left and right hand derivatives to show that the quadratic form Q_h is indefinite and therefore has a null vector.

Step 3 Use the Hahn-Banach theorem in an appropriate Hilbert space consisting of pairs (p, f) where p is a symmetric $(0, 2)$ tensor on M and f a function on ∂M to show that the pair $(0, 1)$ lies in the convex hull of the pairs $(\tau(u), u^2)$ for first eigenfunctions u.

This tells us that there are positive $a_1, \ldots, a_n$ and eigenfunctions $u_1, \ldots, u_n$ so that

$$\sum a_i^2 \tau(u_i) = 0 \text{ on } M, \quad \sum a_i^2 u_i^2 = 1 \text{ on } \partial M.$$

It then follows that the map $\varphi = (a_1 u_1, \ldots, a_n u_n)$ is a (possibly branched) proper minimal immersion in the unit ball $\mathbb{B}^n$. It can then be checked that φ is an isometry on ∂M. □

Theorem 1.4.3 shows that any maximizing metric has a nice geometric characterization as the induced metric on a free boundary minimal surface in a ball. A main difficult question is that of existence of maximizing metrics. This will be discussed in the next section.

1.4.3 Existence of Maximizing Metrics and Sharp Eigenvalue Bounds

A main result of [20] proves existence of a maximizing metric for surfaces of genus zero with any number of boundary components.

Theorem 1.4.4 ([20]) *For any $k \geq 1$ there exists a smooth metric g on the surface M of genus 0 with k boundary components such that $\sigma_1(g)L_g(\partial M) = \sigma^*(0, k)$.*

We refer the reader to [20, Theorem 5.6] for the full details of the proof. Here we indicate only some key ingredients that go into the proof. Very roughly the proof involves first controlling the conformal structure of metrics near the supremum, and then controlling the metrics themselves.

Part 1 We prove boundedness of the conformal structure for nearly maximizing metrics for surfaces of genus 0 with arbitrarily many boundary components.

This involves showing that the supremum value for $\sigma_1 L$ strictly increases when a boundary component is added. Specifically, we show that if $\sigma^*(0, k - 1)$ is achieved, then $\sigma^*(0, k) > \sigma^*(0, k - 1)$. It is then shown by delicate constructions of comparison functions that if the conformal structure degenerates for a sequence, then σ_1 for the sequence must be asymptotically bounded above by the supremum for surfaces with fewer boundary components. It follows that the conformal structure does not degenerate for a maximizing sequence.

There are two general ways in which metrics can degenerate in the problem. The first is that the conformal structure might degenerate and the second is that the boundary arclength measure might become singular even though the conformal class is controlled.

Part 2 We show that for *any* compact surface M with boundary (not necessarily genus zero), a smooth maximizing metric g exists on M provided the conformal structure is controlled for any metric near the maximum.

To prove this we use a canonical regularization procedure to produce a special maximizing sequence and take a weak* limit of the boundary measures. We then prove that first eigenfunctions give a branched conformal minimal immersion into the ball which is freely stationary. The regularity at the boundary then follows. Finally we observe that the metric on the boundary can be recovered from the map and it is smooth. Branch points do not occur on the boundary.

There has been recent progress on existence of maximizing metrics for surfaces of genus $\gamma > 0$ [31, 38].

In a few cases we are able to explicitly characterize the maximizing metric, and obtain sharp eigenvalue bounds. We know $\sigma^*(0, 1) = 2\pi$. The next result identifies $\sigma^*(0, 2)$.

Theorem 1.4.5 ([20]) *For any metric g on the annulus M we have*

$$\sigma_1(g)L_g(\partial M) \leq (\sigma_1 L)_{cc}$$

with equality if and only if (M, g) is σ-homothetic to the critical catenoid. In particular,

$$\sigma^*(0, 2) = (\sigma_1 L)_{cc} = \frac{4\pi}{T_0}$$

where $T_0 \approx 1.2$ is the unique positive solution of $\coth t = t$.

Remark 1.4.2 Observe that the coarse upper bound, Theorem 1.4.2, for $\gamma = 0$ and $k = 2$ gives an upper bound of 4π for the annulus. By Theorem 1.4.5, the sharp upper bound for the annulus is $\sigma^*(0, 2) = (\sigma_1 L)_{cc} = 4\pi/T_0 \approx 4\pi/1.2$.

Outline of Proof The proof is a consequence of three results:

- By Theorem 1.4.4 there exists a metric g on the annulus such that $\sigma_1(g)L_g(\partial M) = \sigma^*(0, 2)$.
- By the characterization of maximizing metrics, Theorem 1.4.3, g is σ-homothetic to the induced metric from a free boundary minimal immersion of the annulus M into the ball by first Steklov eigenfunctions.
- We show that the critical catenoid is the unique free boundary minimal annulus in $\mathbb{B}^n$ such that the coordinate functions are *first* Steklov eigenfunctions (see Theorem 1.5.3 discussed in Sect. 1.5.2). □

We also show that the conformal structure is controlled for nearly maximizing metrics on the Möbius band. It then follows from Part 2 of the proof of Theorem 1.4.4 that there exists a metric that maximizes the first nonzero normalized eigenvalue on the Möbius band. We explicitly and uniquely (up to σ-homothety) characterize the maximizing metric as the induced metric on the *critical Möbius band*.

Example 1.4.6 (Critical Möbius Band) We think of the Möbius band M as $\mathbb{R} \times \mathbb{S}^1$ with the identification $(t, \theta) \approx (-t, \theta + \pi)$. There is a minimal embedding of M into $\mathbb{R}^4$ given by

$$\varphi(t, \theta) = (2 \sinh t \cos \theta, 2 \sinh t \sin \theta, \cosh 2t \cos 2\theta, \cosh 2t \sin 2\theta).$$

For a unique choice of T_0 the restriction of φ to $[-T_0, T_0] \times \mathbb{S}^1$ is an embedding into a ball meeting the boundary orthogonally. Explicitly T_0 is the unique positive solution of $\coth t = 2 \tanh 2t$. We may rescale the radius of the ball to 1 to get the *critical Möbius band*.

Theorem 1.4.6 ([20]) *For any metric g on the Möbius band M we have*

$$\sigma_1(g) L_g(\partial M) \le (\sigma_1 L)_{cmb} = 2\sqrt{3}\pi$$

with equality if and only if g is σ-homothetic to the induced metric on the critical Möbius band.

As in the proof of Theorem 1.4.5, the proof of Theorem 1.4.6 uses a minimal surface uniqueness theorem showing that the critical Möbius band is the unique free boundary minimal Möbius band in $\mathbb{B}^n$ such that the coordinate functions are *first* Steklov eigenfunctions [20, Theorem 7.4].

While in general we cannot expect to be able to explicitly characterize the maximizing metrics, for surfaces of genus zero with any number of boundary components we show that maximizing metrics arise from free boundary surfaces in $\mathbb{B}^3$ that are embedded and star-shaped with respect to the origin.

Theorem 1.4.7 ([20]) *The sequence $\sigma^*(0, k)$ is strictly increasing in k and for each k, $\sigma^*(0, k)$ is achieved by the induced metric on an embedded free boundary minimal surface Σ_k in $\mathbb{B}^3$.*

In particular, as a consequence of the results on the eigenvalue problem we obtain existence of embedded free boundary minimal surfaces in $\mathbb{B}^3$ of genus zero with any number $k > 0$ of boundary components. Previously the only known free boundary minimal surfaces in $\mathbb{B}^3$ were the equatorial plane disk and the critical catenoid.

Outline of Proof of Theorem 1.4.7 By Theorems 1.4.4 and 1.4.3, $\sigma^*(0, k)$ is achieved by the induced metric on a free boundary minimal surface Σ_k in $\mathbb{B}^n$ immersed by first eigenfunctions. For surfaces of genus zero, the multiplicity of σ_1 is at most 3, and so $n = 3$. The restrictions of the linear functions are first eigenfunctions, and so by the nodal domain theorem have no critical points on their zero set. Using this we show that Σ_k does not contain the origin and is embedded and star-shaped (see [20, Theorem 8.3] for details).   □

1.4.4   Higher Dimensional Problems

For Riemannian manifolds of dimension $n \geq 3$ in general there is no upper bound on the first normalized Steklov eigenvalue that is independent of the metric [8], in contrast with the case of surfaces. On the other hand, there is an upper bound for domains in $\mathbb{R}^n$. In fact, for any bounded domain $\Omega \subset \mathbb{R}^n$ there are upper bounds on all normalized Steklov eigenvalues, $\sigma_j(\Omega)|\partial\Omega|^{\frac{1}{n-1}} \leq C(n)j^{\frac{2}{n}}$ [9]. For the first eigenvalue, an explicit upper bound can be directly obtained as follows. Using inverse stereographic projection and a standard balancing argument, there exists a conformal map $F : \Omega \to \mathbb{S}^n \subset \mathbb{R}^{n+1}$ with $\int_{\partial\Omega} F = 0$. Then, using the component functions $F_i, i = 1, \ldots, n + 1$, as test functions in the variational characterization of σ_1 (cf. Theorem 1.4.1), one obtains the bound:

$$\overline{\sigma}_1(\Omega) := \sigma_1(\Omega)|\partial\Omega|^{\frac{1}{n-1}} \leq \frac{n^{\frac{1}{n-1}}|\mathbb{S}^n|^{\frac{2}{n}}}{|\mathbb{B}^n|^{\frac{n-2}{n(n-1)}}}.$$

See [20, Proposition 2.1] for details. This upper bound is unlikely to be sharp, and leaves open the question of finding the sharp value for the upper bound.

Question 1.4.1 ([24]) On which domain (or in the limit of which sequence of domains) is the supremum of $\overline{\sigma}_1(\Omega)$ over all bounded domains $\Omega \subset \mathbb{R}^n$ realized?

If we normalize the volume of Ω rather than the volume of the boundary $\partial\Omega$, a theorem of Brock [4] shows that a round ball maximizes σ_1 among all smooth domains with the same (or larger) volume.

Theorem 1.4.8 ([4]) *For any bounded domain Ω in $\mathbb{R}^n$,*

$$\sigma_1(\Omega) \leq \sigma_1(\Omega^*)$$

where Ω^* is a ball with the same volume as Ω, $|\Omega^*| = |\Omega|$. Equality holds if and only if Ω is a ball.

Recall that Weinstock's Theorem (Theorem 1.4.1) says that the round disk uniquely maximizes σ_1 among all simply connected domains in the plane with the same boundary length: $\sigma_1 L \leq 2\pi$. For $n = 2$, Brock's inequality says that $\sigma_1\sqrt{A} \leq \sqrt{\pi}$. By the isoperimetric inequality, for any domain in the plane we have $\sqrt{A} \leq L/(2\sqrt{\pi})$, and so Weinstock's theorem implies Brock's theorem for simply connected plane domains. On the other hand Brock's theorem holds for arbitrary plane domains and domains in $\mathbb{R}^n$ for $n \geq 3$. This leads to the question of whether there is an analogue to Weinstock's theorem in higher dimensions.

It was recently proved that Weinstock's inequality holds in any dimension, provided we restrict to the class of convex sets:

Theorem 1.4.9 ([5]) *For every bounded convex set $\Omega \subset \mathbb{R}^n$,*

$$\sigma_1(\Omega) \leq \sigma_1(\Omega^*)$$

where Ω^* is a ball with the same boundary volume as Ω, $|\partial \Omega^*| = |\partial \Omega|$. Equality holds if and only if Ω is a ball.

Since Weinstock's inequality holds for any simply-connected domain in the plane, one may ask whether the ball maximizes σ_1 more generally among *contractible* domains in $\mathbb{R}^n$ with the same (or larger) boundary volume. The isoperimetric inequality implies that any domain with the same volume has larger boundary volume, so this would imply Brock's theorem for contractible domains. However, in [21] it is shown that the analogue of Weinstock's inequality is not true for $n \geq 3$.

Theorem 1.4.10 ([21]) *For $n \geq 3$ there exist smooth contractible domains $\Omega \subset \mathbb{R}^n$ with $|\partial \Omega| = |\partial \mathbb{B}^n|$ but $\sigma_1(\Omega) > \sigma_1(\mathbb{B}^n)$.*

In particular, although it is true that for domains in $\mathbb{R}^n$ with fixed boundary volume there is an upper bound on σ_1, the upper bound is not achieved by the ball.

Idea of Proof First consider the annular domain $\mathbb{B}_1 \setminus \mathbb{B}_\epsilon$, $0 < \epsilon \ll 1$ (Fig. 1.2(a)). By explicit computation using separation of variables,

$$\sigma_1(\mathbb{B}_1 \setminus \mathbb{B}_\epsilon) = 1 - \frac{n}{n-1}\epsilon^n + O(\epsilon^{n+1}).$$

On the other hand, $|\partial(\mathbb{B}_1 \setminus \mathbb{B}_\epsilon)| = (1 + \epsilon^{n-1})|\partial \mathbb{B}_1|$, and so for ϵ small,

$$\sigma_1(\mathbb{B}_1 \setminus \mathbb{B}_\epsilon)|\partial(\mathbb{B}_1 \setminus \mathbb{B}_\epsilon)|^{\frac{1}{n-1}} > \sigma_1(\mathbb{B}_1)|\partial \mathbb{B}_1|^{\frac{1}{n-1}}.$$

This holds for any $n \geq 2$ (see Example 1.4.3 when $n = 2$).

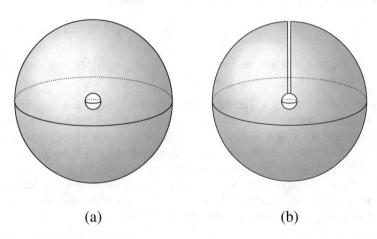

(a)   (b)

Fig. 1.2 (**a**) Annular domain $\mathbb{B}_1 \setminus \mathbb{B}_\epsilon$, $0 < \epsilon \ll 1$. (**b**) A rough picture of a contractible domain with larger $\bar{\sigma}_1$ than the ball

For $n \geq 3$, we can modify the annular domain $\mathbb{B}_1 \setminus \mathbb{B}_\epsilon$ to make it contractible while changing the normalized eigenvalue by an arbitrarily small amount. This is accomplished by connecting the boundary components by a thin tube T_δ (Fig. 1.2(b)), and showing that the construction can be done while keeping the normalized eigenvalue nearly unchanged. A key technical difficulty is to prove that for a sequence of eigenfunctions, the L^2-norm on the boundary doesn't concentrate on the boundary of the tube T_δ as the radius of the tube $\delta \to 0$ [21, Lemma 4.2].   □

This construction leads to a more general question about boundary connectedness. Recall that for $n = 2$ we showed that adding boundary components increases the value of σ_1 normalized by boundary length (see Part 1 of outline of proof of Theorem 1.4.4). One is then led to ask whether a similar phenomenon is true in higher dimensions. In [21] it is shown that this also fails for $n \geq 3$.

Theorem 1.4.11 ([21]) *Given any compact Riemannian manifold Ω^n with non-empty boundary and $n \geq 3$, and given any $\epsilon > 0$ there exists a smooth subdomain Ω_ϵ of Ω with connected boundary such that*

$$|\Omega| - |\Omega_\epsilon| < \epsilon, \ ||\partial\Omega| - |\partial\Omega_\epsilon|| < \epsilon, \ and \ |\sigma_1(\Omega) - \sigma_1(\Omega_\epsilon)| < \epsilon.$$

The idea of the proof is similar; we consider the effect of adding thin tubes connecting boundary components.

In summary, we see that some of the refined results that are true for surfaces, don't hold in higher dimensions:

- For $n = 2$, the disk maximizes σ_1 among simply connected plane domains with fixed boundary length. For $n \geq 3$ this is not true. There exist domains in $\mathbb{R}^n$ that are diffeomorphic to the ball and have the same boundary volume as the ball but have larger σ_1.
- For $n = 2$, adding boundary components increases the supremum of $\sigma_1 L$. For $n \geq 3$ this is not true. The number of boundary components does not affect the supremum of the first normalized eigenvalue.

1.5 Free Boundary Minimal Surfaces in $\mathbb{B}^n$

As we have seen, there is a close connection between free boundary minimal surfaces in $\mathbb{B}^n$ and the Steklov eigenvalue problem. In this section we give an overview of results on existence, uniqueness, and Morse index of free boundary minimal surfaces in $\mathbb{B}^n$. A number of the results we will discuss in this section are either a consequence of, or have interesting applications to, the Steklov problem. We refer the reader also to the survey article [36].

1.5.1   Existence of Free Boundary Minimal Surfaces in $\mathbb{B}^3$

As a consequence of the results of [20] on the Steklov eigenvalue problem, discussed in Sects. 1.4.2 and 1.4.3, we have the following minimal surface existence theorem:

Theorem 1.5.1 ([20]) *For every $k \geq 1$ there is an embedded free boundary minimal surface in $\mathbb{B}^3$ of genus 0 with k boundary components. Moreover, these surfaces are embedded by first Steklov eigenfunctions.*

Previously, the only known free boundary minimal surfaces in $\mathbb{B}^3$ were the equatorial plane disk and the critical catenoid. Theorem 1.5.1 and the connection between free boundary minimal surfaces in a ball and the Steklov eigenvalue problem have raised the interest of finding more examples of properly embedded minimal surfaces in the unit ball.

We note that there are many parallels between free boundary minimal submanifolds in $\mathbb{B}^n$ and closed minimal submanifolds in $\mathbb{S}^n$ (e.g. see Sect. 1.2). In the closed case, Lawson [35] proved that a closed orientable surface of any genus can be realized as an embedded minimal surface in $\mathbb{S}^3$, while any non-orientable closed surface except $\mathbb{RP}^2$ can be realized as a minimal immersion into $\mathbb{S}^3$. A central question for free boundary minimal surfaces in the ball is:

Question 1.5.1 Which compact orientable surfaces with boundary can be realized as properly embedded free boundary minimal surfaces in $\mathbb{B}^3$?

By Theorem 1.5.1, we know that any genus zero surface can be minimally embedded into $\mathbb{B}^3$ as a free boundary solution. Existence of maximizing metrics for surfaces of genus $\gamma > 0$ with $k > 0$ boundary components would give existence of a (possibly branched) free boundary minimal surface in $\mathbb{B}^n$, for *some $n \geq 3$*, but not necessarily in $\mathbb{B}^3$. Recently, further new embedded free boundary minimal surfaces in $\mathbb{B}^3$ have been constructed using gluing techniques and min–max constructions. We outline the currently known examples:

- $\gamma = 0$, $k \geq 1$: Theorem 1.5.1 gives existence of embedded free boundary minimal surfaces in $\mathbb{B}^3$ of genus 0 with any number $k \geq 1$ of boundary components, as a consequence of the existence of a metric that maximizes the first normalized Steklov eigenvalue on any surface of genus zero (Theorem 1.4.4) and the characterization of maximizing metrics (Theorem 1.4.3). Folha et al. [15] gave an independent construction for $\gamma = 0$ when is k *large*, by doubling the disk. The surfaces of [15] are invariant under a group of reflections.
- $\gamma = 1$, k *large*: By doubling the disk and gluing in a catenoidal neck in the center, [15] also constructs embedded free boundary minimal surfaces in $\mathbb{B}^3$ of genus 1 with any sufficiently large number k of boundary components.
- γ *large*, $k = 3$: By gluing the critical catenoid and the equatorial plane disk, Kapouleas and Martin Li [28] constructed embedded free boundary minimal surfaces in $\mathbb{B}^3$ with three boundary components and any sufficiently large genus.

Such surfaces were independently constructed by Ketover [33] using equivariant min–max theory.

- $\gamma \geq 0$, $k = 1$: By tripling the equatorial disk, Kapouleas and Wiygul [29] constructed embedded free boundary minimal surfaces in $\mathbb{B}^3$ with connected boundary and any sufficiently *large* genus. Very recently, Carlotto et al. [7] have used min–max theory to construct surfaces with connected boundary and *arbitrary* genus.

These are currently, as far as we are aware, the only known examples of free boundary minimal surfaces in $\mathbb{B}^3$. For surfaces of other topological types, the general existence Question 1.5.1 remains open.

1.5.2   Uniqueness of Free Boundary Minimal Surfaces in $\mathbb{B}^n$

As we have seen in Sect. 1.5.1, there are relatively few known examples of free boundary minimal surfaces in $\mathbb{B}^3$. In view of this, another question one may ask is:

Question 1.5.2 Given a compact orientable surface with boundary, in how many ways can one realize it as a properly embedded free boundary minimal surface in the unit ball $\mathbb{B}^n$?

A classical result of J.C.C. Nitsche [43] shows that the only free boundary minimal disk in $\mathbb{B}^3$ is the equatorial plane disk. It was expected that in higher codimension there should exist other free boundary minimal disks. However in [19] we proved that uniqueness also holds in higher codimension: any free boundary minimal disk in $\mathbb{B}^n$ is an equatorial plane disk. This result is surprising by analogy with the case of minimal submanifolds of spheres: a result of Almgren [1] shows that a minimal $\mathbb{S}^2$ in $\mathbb{S}^3$ is totally geodesic, however there are many minimal immersions of $\mathbb{S}^2$ in $\mathbb{S}^n$ for $n \geq 4$ that are not totally geodesic. Our result [19] shows that one may expect more rigidity for free boundary minimal surfaces than for the analogous case of minimal surfaces in spheres. In [19] we also considered more generally the case of free boundary disks with parallel mean curvature in a ball in a space of constant curvature (generalizing the results of [43, 48] to higher codimension).

Theorem 1.5.2 ([19]) *Let $u : \mathbb{D} \to B$ be a proper branched immersion with parallel mean curvature vector, where B is a geodesic ball in an n-dimensional space of constant curvature, such that $u(\mathbb{D})$ meets ∂B orthogonally. Then Σ is contained in a three-dimensional totally geodesic submanifold, and is totally umbilic.*

The method of proof is complex analytic as in the case $n = 3$. Here we outline the idea of the proofs in the case of free boundary minimal disks in $\mathbb{B}^n$.

In general, suppose Σ is a free boundary minimal surface in $\mathbb{B}^n$. We may parametrize Σ by a conformal harmonic map $u : M \to \mathbb{B}^n$ from a surface M. Let $z = x_1 + i x_2$ be local complex coordinates on M. The condition that u is harmonic is expressed by the equation $u_{z\bar{z}} = 0$, and the condition that u is conformal is expressed by the equation $u_z \cdot u_z = 0$.

Case $n = 3$ Consider the Hopf differential $\Phi(z) = \varphi(z)dz^2$, where $\varphi(z) = u_{zz} \cdot N$ and N is a unit normal to Σ. We have

$$\varphi(z) = \frac{1}{4}[(h_{11} - h_{22}) - 2i\, h_{12}],$$

where $h_{ij} = \nabla_{\frac{\partial}{\partial x_i}} \frac{\partial u}{\partial x_j} \cdot N$, $i, j = 1, 2$ are the components of the scalar second fundamental form of Σ relative to the local coordinates. Also,

$$(u_{zz} \cdot N)_{\bar{z}} = (u_{z\bar{z}})_z \cdot N + u_{zz}^{\perp} \cdot \nabla_{\frac{\partial}{\partial \bar{z}}} N = 0,$$

since $u_{z\bar{z}} = 0$ and N is parallel in the normal bundle, where $u_{zz}^{\perp}$ denotes the component of u_{zz} orthogonal to $\Sigma = u(M)$. Thus Φ is a holomorphic quadratic differential on M. The free boundary condition implies that Φ is real on ∂M. To see this, choose local complex coordinates $z = x_1 + ix_2$ near a boundary point p such that at p, $\frac{\partial u}{\partial x_1}$ is a unit tangent T to $\partial \Sigma$ and $\frac{\partial u}{\partial x_2} = \eta$, where η is the outward unit conormal of Σ. By the free boundary condition, η is the outward unit normal to $\partial \mathbb{B}^n$, which is the position vector X in $\mathbb{R}^n$. We have $dz(T) = 1$ at p, and

$$\Phi(T) = \varphi(p).$$

By the free boundary condition, at a boundary point the unit tangent T and unit normal η form a principal basis, since

$$h_{12} = (\nabla_T \eta) \cdot N = (\nabla_T X) \cdot N = T \cdot N = 0.$$

Therefore, $\Phi(T) = \varphi(p) = \frac{1}{4}(h_{11} - h_{22})$ is real. Thus Φ is holomorphic on M and real on ∂M.

We now specialize to the case where $M = \mathbb{D}$ is the disk. In polar coordinates $z = re^{i\theta}$, $dz(\frac{\partial}{\partial \theta}) = iz$, and $\Phi(\frac{\partial}{\partial \theta}) = -z^2\varphi(z)$. Therefore, $z^2\varphi(z)$ is holomorphic and real on $\partial \mathbb{D}$, and so $z^2\varphi(z)$ is constant. Since $z^2\varphi(z)$ vanishes at the origin, $z^2\varphi(z) \equiv 0$, and so $\varphi(z) \equiv 0$. Therefore $h_{11} = h_{22}$ and $h_{12} = 0$. Since Σ is minimal, $h_{11} + h_{22} = 0$. Hence $h_{ij} = 0$, $i, j = 1, 2$ and Σ is a flat disk.

Case $n \geq 3$ In higher codimension, a difficulty is that the second fundamental form is vector valued. That is, the corresponding differential $u_{zz}^{\perp}dz^2$ has values in the normal bundle. However, there is an associated quartic differential $(u_{zz}^{\perp} \cdot u_{zz}^{\perp})\, dz^4$ gotten by squaring the Hopf differential using the inner product (cf. [6]). First observe that the minimality of Σ implies that $u_{zz}^{\perp}$ is a holomorphic section of the normal bundle of Σ, as follows. The normal bundle is smooth across branch points, and

$$u_{zz}^{\perp} = u_{zz} - \frac{u_{zz} \cdot u_z}{|u_{\bar{z}}|^2} u_{\bar{z}} - \frac{u_{zz} \cdot u_{\bar{z}}}{|u_z|^2} u_z = u_{zz} - \frac{u_{zz} \cdot u_{\bar{z}}}{|u_z|^2} u_z,$$

where in the second equality we used that

$$u_{zz} \cdot u_z = \frac{1}{2}(u_z \cdot u_z)_z = 0,$$

since u is conformal. Then,

$$\nabla_{\frac{\partial}{\partial \bar{z}}} u_{zz}^{\perp} = (u_{z\bar{z}})_z - \frac{u_{zz} \cdot u_{\bar{z}}}{|u_z|^2} u_{z\bar{z}} + \text{tangential terms} = \text{tangential terms},$$

where in the second equality we used that $u_{z\bar{z}} = 0$, since u is harmonic. Therefore,

$$\nabla_{\frac{\partial}{\partial \bar{z}}}^{\perp}\left(u_{zz}^{\perp}\right) = 0. \tag{1.3}$$

Let $\varphi(z) = u_{zz}^{\perp} \cdot u_{zz}^{\perp}$. Then $\varphi_{\bar{z}} = 2 u_{zz}^{\perp} \cdot \nabla_{\bar{z}}(u_{zz}^{\perp}) = 0$, by (1.3). Therefore, $\Phi(z) = \varphi(z)\,dz^4$ is a holomorphic quartic differential. We have

$$u_{zz}^{\perp} = \nabla_{\frac{\partial}{\partial \bar{z}}}^{\perp} u_z = \frac{1}{4}\left[(h_{11} - h_{22}) - 2ih_{12}\right] = \frac{1}{2}(h_{11} - ih_{12}),$$

where here $h_{ij} = \nabla_{\frac{\partial}{\partial x_i}}^{\perp} \frac{\partial u}{\partial x_j}, i,\ j = 1,\ 2$, denote the components of the vector valued second fundamental form relative to the local coordinates, and in the second equality we used the minimality of Σ, $h_{11} + h_{22} = 0$. Then,

$$\varphi(z) = u_{zz}^{\perp} \cdot u_{zz}^{\perp} = \frac{1}{4}\left(|h_{11}|^2 - |h_{12}|^2 - 2ih_{11} \cdot h_{12}\right).$$

As in the $n = 3$ case, the free boundary condition implies that Φ is real on ∂M. To see this, choose local complex coordinates $z = x_1 + ix_2$ near a boundary point p such that at p, $\frac{\partial u}{\partial x_1}$ is a unit tangent T to $\partial\Sigma$ and $\frac{\partial u}{\partial x_2} = \eta = X$. Then

$$h_{12} = \nabla_T^{\perp}\eta = (\nabla_T X)^{\perp} = T^{\perp} = 0.$$

We have $dz(T) = 1$ at p, and

$$\Phi(T) = \varphi(p) = \frac{1}{4}|h_{11}|^2.$$

Thus Φ is holomorphic on M and real on ∂M.

We now specialize to the case where $M = \mathbb{D}$ is the disk. In polar coordinates $z = re^{i\theta}$, $dz(\frac{\partial}{\partial\theta}) = iz$, and $\Phi(\frac{\partial}{\partial\theta}) = z^4\varphi(z)$. Therefore, $z^4\varphi(z)$ is holomorphic and real on $\partial\mathbb{D}$, and so $z^4\varphi(z)$ is constant. Since $z^4\varphi(z)$ vanishes at the origin, $z^4\varphi(z) \equiv 0$, and so $\varphi(z) \equiv 0$. Notice that unlike the $n = 3$ case, this does not immediately imply that the second fundamental form of Σ is zero, since the expression for φ is more complicated in this case. However, at a boundary point p, in the adapted basis T, η as above, we have $h_{12} = 0$ and so $h_{11} = 0$. Therefore, the second fundamental form of Σ vanishes on $\partial\Sigma$. Notice that,

$$0 = h_{11} = (\nabla_T T)^\perp = k$$

where k is the curvature of $\partial\Sigma$ in the sphere $\partial\mathbb{B}^n$, since by the free boundary condition the component of a vector orthogonal to Σ at a boundary point is tangent to the sphere. Therefore, $\partial\Sigma$ is a great circle. Finally, by the maximum principle, Σ must be a flat disk. $\square$

What about free boundary surfaces of other topological types? The next natural case to consider after the disk is the annulus, and the only known free boundary minimal annulus in $\mathbb{B}^n$ is the critical catenoid in $\mathbb{B}^3$. We note that the Hopf differential argument outlined above gives some information about free boundary annuli. Any metric on the annulus is conformal to the product metric on a cylinder $A_T := [-T, T] \times \mathbb{S}^1$ for some $T > 0$. If Σ is a free boundary minimal annulus in $\mathbb{B}^n$, we can parametrize Σ by a conformal harmonic map

$$u : [-T, T] \times \mathbb{S}^1 \to \mathbb{B}^n.$$

Consider complex coordinates $z = t + i\theta$ on the cylinder A_T. The arguments given above imply that $\varphi(z) = c$ is a real constant on the annulus. In particular, $h_{11} = c$ and $h_{12} = 0$ in these coordinates. That is, (t, θ) are principal coordinates. Since $\frac{\partial}{\partial t}$ is not a unit vector, this does not say that the principal curvature k is constant, but that $k = c/|\frac{\partial u}{\partial t}|^2$.

While it is expected that there should exist other *immersed* free boundary annuli in $\mathbb{B}^3$, the critical catenoid is expected to be the only *embedded* free boundary annulus in $\mathbb{B}^3$.

Conjecture 1.5.1 The critical catenoid is the unique embedded free boundary minimal surface in $\mathbb{B}^3$ that is homeomorphic to an annulus.

This may be viewed as a free boundary analogue of the Lawson conjecture on the uniqueness of the Clifford torus in $\mathbb{S}^3$. Although the Lawson conjecture was solved [3], the proof does not readily extend to the free boundary setting because of problems with boundary terms. Motivated by our work on the Steklov eigenvalue problem, we proved the following uniqueness theorem for the critical catenoid.

Theorem 1.5.3 ([20]) *Assume that Σ is a free boundary minimal annulus in $\mathbb{B}^n$ such that the coordinate functions are first Steklov eigenfunctions. Then $n = 3$ and Σ is the critical catenoid.*

Recall that any free boundary minimal surface in $\mathbb{B}^n$ is characterized by the condition that the coordinate functions are Steklov eigenfunctions with eigenvalue 1. Theorem 1.5.3 characterizes the critical catenoid as the only free boundary minimal annulus in $\mathbb{B}^n$ such that coordinate functions are *first* Steklov eigenfunctions; that is, such that $\sigma_1(\Sigma) = 1$. Theorem 1.5.3 has several interesting applications:

(i) **Sharp eigenvalue bound on the annulus**: Theorem 1.5.3 is used in the proof of Theorem 1.4.5 to characterize metrics that maximize the first Steklov eigenvalue on the annulus, and prove a sharp upper bound on the first normalized Steklov eigenvalue among all smooth metrics on the annulus.

(ii) **Strong rigidity of the critical catenoid in $\mathbb{B}^n$**: Theorem 1.5.3 implies that *any free boundary minimal annulus in $\mathbb{B}^n$ that is (C^2-) close to the critical catenoid is a rotation of the critical catenoid* [22, Theorem 3.3].

(iii) **Index of free boundary minimal annuli**: Theorem 1.5.3 implies that *the critical catenoid is the only free boundary minimal annulus of index 4*. This will be discussed in Sect. 1.5.3.

Theorem 1.5.3 has also been used by McGrath [39] to prove that *the critical catenoid is the only embedded free boundary annulus in $\mathbb{B}^3$ that is invariant under reflection through the coordinate planes*, providing further evidence toward Conjecture 1.5.1. More generally, McGrath [39] gives general criteria which imply that an embedded free boundary minimal surface Σ in $\mathbb{B}^3$ that is invariant under a group of reflections has $\sigma_1(\Sigma) = 1$.

For a free boundary minimal submanifold in the unit ball it is always true that 1 is a Steklov eigenvalue and the coordinate functions are corresponding eigenfunctions. It is a subtle question to determine if 1 is the *first* Steklov eigenvalue. For embedded solutions of codimension one it is natural to conjecture that 1 is the first eigenvalue.

Conjecture 1.5.2 Let Σ be a compact properly embedded free boundary minimal hypersurface in $\mathbb{B}^n$. Then $\sigma_1(\Sigma) = 1$.

This conjecture would imply Yau's Conjecture for minimal hypersurfaces in spheres because it could be applied to the cone. With M. Li we proved in [16, Theorem 3.1] that $\sigma_1(\Sigma) > 1/2$. A resolution of Conjecture 1.5.2 would have many applications including a unified proof of uniqueness of the critical catenoid as the only embedded free boundary annulus in $\mathbb{B}^3$ and a proof of the uniqueness of the Clifford torus (see [3]) as the only embedded minimal torus in $\mathbb{S}^3$. The results of [39] show that most of the known free boundary minimal surfaces in $\mathbb{B}^3$ are embedded by first eigenfunctions, providing evidence toward Conjecture 1.5.2 when $n = 3$.

We conclude this section by giving an outline of the proof of Theorem 1.5.3. One of the complications in proving the theorem is that the effect of conformal transformations on the volume of free boundary minimal submanifolds in $\mathbb{B}^n$ is not understood. Specifically, one would like to know:

Question 1.5.3 If Σ is a k-dimensional free boundary minimal submanifold in $\mathbb{B}^n$, and $f : \mathbb{B}^n \to \mathbb{B}^n$ is conformal, then is $|f(\Sigma)| \leq |\Sigma|$?

It is true that if Σ is a closed k-dimensional minimal submanifold of $\mathbb{S}^n$ and $f : \mathbb{S}^n \to \mathbb{S}^n$ is conformal, then $|f(\Sigma)| \leq |\Sigma|$ [13, 37]. This result in the two-dimensional case plays a key role in the characterization of the maximizing metrics for λ_1 on the torus and Klein bottle (see [27, 40, 41]). If Question 1.5.3 were true, it would imply the uniqueness Theorem 1.5.3, since it would imply that any conformal automorphism of Σ is an isometry (as in [40, Theorem 1], [13, 3.3 Corollaire]). Since any metric on the annulus is conformal to a rotationally symmetric metric, the result would then follow from the analysis of rotationally symmetric metrics on the annulus in [17, Section 3]. Unfortunately, we were not able to show that Question 1.5.3 is true. Furthermore it is not clear whether the volume even decreases to second order. We refer the reader to Section 3 of [18] for further discussion of volumes of free boundary minimal submanifolds in $\mathbb{B}^n$. By a subtle argument, we were able to show that the second variation of volume is nonpositive for the vector field $v^\perp$, for any $v \in \mathbb{R}^n$:

Theorem 1.5.4 *If Σ^k is a free boundary minimal submanifold in $\mathbb{B}^n$ and $v \in \mathbb{R}^n$, then we have*

$$\delta^2 \Sigma(v^\perp, v^\perp) = -k \int_\Sigma |v^\perp|^2 \, d\mu.$$

If Σ is not contained in a product $\Sigma_0 \times \mathbb{R}$ where Σ_0 is a free boundary solution, then the Morse index of Σ is at least n. In particular, if $k = 2$ and Σ is not a plane disk, its index is at least n.

Theorem 1.5.4 plays a key role in the proof of Theorem 1.5.3.

Outline and Ideas Behind the Proof of Theorem 1.5.3 Assume Σ is free boundary minimal annulus in $\mathbb{B}^n$ such that the coordinate functions are first Steklov eigenfunctions; i.e. $\sigma_1(\Sigma) = 1$. A multiplicity bound implies that $n = 3$. We may parametrize Σ by a proper conformal harmonic map

$$u : M \to \mathbb{B}^3$$

where $M = [-T, T] \times \mathbb{S}^1$ with coordinates (t, θ). The main idea is to show that u is $\mathbb{S}^1$-invariant. Consider the conformal Killing vector field $X = \frac{\partial u}{\partial \theta}$ along Σ. $\square$

Main Idea Show that the three components of X are first eigenfunctions.

If the components of $X = u_\theta$ were first eigenfunctions, then

$$(u_\theta)_t = \sigma_1 \lambda u_\theta \quad \text{on } \partial\Sigma$$

where $\lambda = |u_t| = |u_\theta|$. Taking the dot product with u_t gives $u_t \cdot u_{\theta t} = 0$, or $\frac{1}{2}(|u_t|^2)_\theta = 0$, and so $\lambda \equiv c$ on $\partial\Sigma$. This implies that Σ is σ-homothetic to a rotationally symmetric annulus. Then by the rotationally symmetric analysis of [17, Section 3], Σ must be σ-homothetic to the critical catenoid. In particular, the conformal structure is the same, and so the T is the same. Since the coordinate functions of Σ are first eigenfunctions, they must then be linear combinations of $\cosh(t)\cos(\theta)$, $\cosh(t)\sin(\theta)$, t. Therefore Σ is a linear image of the critical catenoid, from which it follows that Σ must be a rotation of the critical catenoid. Hence Σ is isometric to the critical catenoid.

Key Observation Connection between the assumption that $\sigma_1(\Sigma) = 1$ and the second variation of energy of the harmonic map u.

First recall the formula for the second variation of energy of the harmonic map $u : M \to \mathbb{B}^3$. If $u_t : M \to \mathbb{B}^3$, $t \in (-\epsilon, \epsilon)$, with $u_0 = u$ and $W = \dot{u}_0$, then

$$\frac{d^2}{dt^2}\bigg|_{t=0} E(u_t) = \int_M |\nabla W|^2 \, da - \int_{\partial M} |W|^2 \, ds := Q(W, W).$$

By assumption, the first Steklov eigenvalue of Σ is 1, $\sigma_1(\Sigma) = 1$. By the variational characterization of σ_1, this implies that for any vector field $W = (W_1, W_2, W_3)$ such that $\int_{\partial\Sigma} W \, ds = 0$,

$$Q(W, W) \geq 0$$

$$= 0 \quad \text{only if } W_1, \ W_2, \ W_3 \text{ are first eigenfunctions}$$

One can check that $Q(X, Y) = 0$ for any vector field Y that is tangent to $\partial\mathbb{B}^3$ along $\partial\Sigma$. Thus, if we knew that $\int_{\partial\Sigma} X = 0$, it would follow that the components of X are first eigenfunctions. The problem is that we do not know that $\int_{\partial\Sigma} X = 0$.

Strategy Modify X by adding a vector field which is negative for Q.

To overcome the problem above, our strategy was to find a vector field Y such that $Q(Y, Y) \leq 0$ and $\int_{\partial\Sigma}(X - Y) \, ds = 0$. It would then follow that $Q(X - Y, X - Y) \leq 0$. But since $\sigma_1 = 1$ and $\int_{\partial\Sigma}(X - Y) \, ds = 0$ we would also have $Q(X - Y, X - Y) \geq 0$. Therefore $Q(X - Y, X - Y) = 0$ and the components of $X - Y$ must be first eigenfunctions. From this it is then possible to show that the components of X are in fact first eigenfunctions, which completes the proof.

There is an obvious choice to make for Y, namely a conformal Killing vector field in the ball. This amounts to taking a vector $v \in \mathbb{R}^3$ and setting

$$Y(v) = \frac{1 + |x|^2}{2} v - (x \cdot v)x$$

where x is the position vector in $\mathbb{R}^3$. It is easy to check that v can be chosen so that $\int_{\partial\Sigma}(X - Y) \, ds = 0$. Since Y is conformal, the second variation of energy $Q(Y, Y)$ is the second variation of the area functional. If it was true that $Q(Y, Y) \leq 0$, we could complete the proof. Unfortunately we were not able to show this despite much effort, as discussed above.

Step 1 Consider the second variation of area

To get around this difficulty we considered the second variation of area for normal variations φN where N is the unit normal vector of Σ. Note that this variation is tangent to $\partial \mathbb{B}^3$ along the boundary. The second variation form for area is given by:

$$S(\varphi, \varphi) = \int_{\Sigma} (|\nabla \varphi|^2 - |A|^2 \varphi^2) \, da - \int_{\partial \Sigma} \varphi^2 \, ds$$

where A denotes the second fundamental form of Σ. By Theorem 1.5.4, for any $v \in \mathbb{R}^3$

$$S(v \cdot N, v \cdot N) \leq 0.$$

For free boundary solutions in the ball there is a natural Jacobi field given by $x \cdot N$. It is in the nullspace of the second variation form S. Therefore, S is nonpositive on the four-dimensional space

$$C = \text{Span}\{N_1, N_2, N_3, x \cdot N\}.$$

This is not sufficient for the eigenvalue problem because the normal deformation does not preserve the conformal structure of Σ in general, and so we cannot conclude that Q is negative on C.

Step 2 Finding vector fields that have negative second variation of energy

In the previous step we found a space C on which the second variation of area is negative, but these vector fields may not have negative second variation of energy. The way we get around this problem is to consider adding a tangential vector field $Y^\top$ so that

$$Y = Y^\top + \varphi N$$

preserves the conformal structure and is tangent to $\mathbb{S}^2$ along $\partial \Sigma$. For conformal vector fields the second variation of area is equal to the second variation of energy.

Given $\varphi \in C^\infty(\Sigma)$, can we find $Y^\top$ tangential to Σ so that $Y = Y^\top + \varphi N$ is conformal and tangent to $\mathbb{S}^2$ along $\partial \Sigma$? This involves solving a Cauchy-Riemann equation with boundary condition to determine $Y^\top$. This problem is generally not solvable, but has a one dimensional obstruction for its solvability because Σ is an annulus. We then get existence for φ in a three dimensional subspace of the span of $N_1, N_2, N_3, x \cdot N$. Finally we can arrange the resulting conformal vector field Y to satisfy the boundary integral condition $\int_{\partial \Sigma} (X - Y) \, ds = 0$, and we have

$$Q(Y, Y) = S(\varphi, \varphi) \leq 0.$$

We refer the reader to [20, Section 6] for the full details of the proof. $\qquad \square$

We also have a uniqueness theorem for the critical Möbius band:

Theorem 1.5.5 *Assume that Σ is a free boundary minimal Möbius band in $\mathbb{B}^n$ such that the coordinate functions are first eigenfunctions. Then $n = 4$ and Σ is the critical Möbius band.*

The proof is similar to the proof of Theorem 1.5.3. Theorem 1.5.5 is used in the proof of Theorem 1.4.6 to characterize metrics that maximize the first Steklov eigenvalue on the Möbius band, and prove a sharp upper bound on the first normalized Steklov eigenvalue among all smooth metrics on the Möbius band. Theorem 1.5.5 also implies local rigidity of the critical Möbius band, that any free boundary minimal Möbius band in $\mathbb{B}^n$ that is (C^2) close to the critical Möbius band is a rotation of the critical Möbius band [22, Theorem 3.4].

1.5.3 Index of Free Boundary Minimal Hypersurfaces in $\mathbb{B}^n$

As we saw in the previous section, an analysis of the second variation of area (Theorem 1.5.4) plays a key role in the proofs of uniqueness of the critical catenoid and Möbius band (Theorems 1.5.3 and 1.5.5). In this final section we will give an overview of some results on the index of free boundary minimal hypersurfaces in $\mathbb{B}^n$.

In general, suppose Σ is a compact k-dimensional immersed submanifold in an n-dimensional Riemannian manifold M with boundary, with $\partial\Sigma \subset \partial M$. Suppose $F_t : \Sigma \to M$ is a one-parameter family of immersions with $F_t(\partial\Sigma) \subset \partial M$, $t \in (-\varepsilon, \varepsilon)$, with F_0 given by the inclusion $\Sigma \hookrightarrow M$. Then the first variation of volume is

$$\left.\frac{d}{dt}\right|_{t=0} |F_t(\Sigma)| = -\int_\Sigma \langle X, H\rangle \, d\mu_\Sigma + \int_{\partial\Sigma} \langle X, \eta\rangle \, d\mu_{\partial\Sigma}$$

where η is the outward unit conormal vector of $\partial\Sigma$, H is the mean curvature vector of Σ in M, and $X = \left.\frac{dF_t}{dt}\right|_{t=0}$ is the variation field. It follows from the first variation formula that Σ is a critical point of the volume functional among all smooth variations X that are tangent to ∂M along $\partial\Sigma$, if Σ is minimal, that is has zero mean curvature $H = 0$, and Σ meets ∂M orthogonally along $\partial\Sigma$. In this case, we say that Σ is a *free boundary minimal submanifold of M*.

The second variation of volume of a free boundary minimal submanifold for normal variations X is given by the index form

$$S(X, X) := \left.\frac{d^2}{dt^2}\right|_{t=0} |F_t(\Sigma)|$$

$$= \int_\Sigma \left(|\nabla^\perp X|^2 - |A^X|^2 - \langle R(X), X\rangle\right) d\mu_\Sigma + \int_{\partial\Sigma} \langle \nabla_X X, \eta\rangle \, d\mu_{\partial\Sigma}.$$

$$(1.4)$$

The *index* of a free boundary minimal submanifold Σ is the maximal dimension of a space of admissible normal variations on which the index form S is negative definite. Σ is *stable* if it is a local minimum of the volume functional, in the sense that the second variation of volume is nonnegative $S(X, X) \geq 0$ for all admissible variations X; that is, the index of Σ is zero.

In the special case where $M = \mathbb{B}^n$, we have by Theorem 1.5.4 ([20, Theorem 3.1]) that for any $v \in \mathbb{R}^n$,

$$S(v^{\perp}, v^{\perp}) = -k \int_{\Sigma} |v^{\perp}|^2 \, d\mu_{\Sigma}.$$

It follows that the index of a k-dimensional free boundary minimal submanifold Σ in $\mathbb{B}^n$ is at least n if Σ is not contained in a product $\Sigma_0 \times \mathbb{R}$ where Σ_0 is a free boundary solution. In particular, for $k = 2$, the *the index of any free boundary minimal surface in $\mathbb{B}^n$ that is not an equatorial plane disk is at least n*. In this section our focus will be on the special case when Σ is a hypersurface, $k = n - 1$, where more can be said.

If Σ is a two-sided free boundary minimal hypersurface in M then the second variation of volume (1.4) of Σ for normal variations $X = \varphi N$, where N is a unit normal field to Σ and $\varphi \in C^{\infty}(\Sigma)$, becomes

$$S(\varphi, \varphi) = \int_{\Sigma} \left(|\nabla\varphi|^2 - \operatorname{Ric}_M(N, N)\,\varphi^2 - |A|^2\varphi^2 \right) d\mu_{\Sigma}$$

$$+ \int_{\partial\Sigma} \varphi^2 \langle \nabla_N N, \eta \rangle \, d\mu_{\partial\Sigma} \qquad (1.5)$$

$$= -\int_{\Sigma} \varphi\, L\varphi \, d\mu_{\Sigma} + \int_{\partial\Sigma} \left(\frac{\partial\varphi}{\partial\eta} - \varphi\, h^{\partial M}(N, N) \right) \varphi \, d\mu_{\partial\Sigma}$$

where $L = \Delta + \operatorname{Ric}(N, N) + |A|^2$, and $h^{\partial M}$ is the second fundamental form of ∂M with respect to the inward unit normal. It follows immediately from the second variation formula (1.5) that any free boundary minimal hypersurface Σ in a manifold M with nonnegative Ricci curvature and convex boundary ∂M is *unstable*, that is has index at least one, since $S(1, 1) < 0$.

For the remainder of this section we will restrict to the case where $M = \mathbb{B}^n$. For a free boundary minimal hypersurface Σ in $\mathbb{B}^n$, the second variation formula (1.5) simplifies to

$$S(\varphi, \varphi) = \int_{\Sigma} \left(|\nabla\varphi|^2 - |A|^2\varphi^2 \right) d\mu_{\Sigma} - \int_{\partial\Sigma} \varphi^2 \, d\mu_{\partial\Sigma} \qquad (1.6)$$

$$= -\int_{\Sigma} \varphi\, L\varphi \, d\mu_{\Sigma} + \int_{\partial\Sigma} \left(\frac{\partial\varphi}{\partial\eta} - \varphi \right) \varphi \, d\mu_{\partial\Sigma}$$

where $L = \Delta + |A|^2$ is the Jacobi operator. The index of Σ is the maximal dimension of a subspace of $C^{\infty}(\Sigma)$ on which the index form S is negative definite, or equivalently, the number of negative eigenvalues of the Jacobi operator with Robin boundary condition,

$$\begin{cases} L\varphi + \lambda\varphi = 0 & \text{on } \Sigma \\ \frac{\partial\varphi}{\partial\eta} = \varphi & \text{on } \partial\Sigma. \end{cases} \tag{1.7}$$

We now summarize some of the known results about the index of free boundary minimal hypersurfaces in $\mathbb{B}^n$.

- If Σ is an equatorial hyperplane in $\mathbb{B}^n$, then index$(\Sigma) = 1$

Proof As noted above, the index is at least one, since $S(1, 1) < 0$. Suppose the index was greater than or equal to two. Then there would exist a two-dimensional subspace $\mathcal{V}$ of $C^\infty(\Sigma)$ containing the constant functions on which the second variation of volume was negative definite. Let $\varphi \in \mathcal{V}$ be orthogonal to the constant functions, $\int_\Sigma \varphi = 0$. However any flat equatorial hyperplane in $\mathbb{B}^n$ minimizes volume subject to the constraint that it separates the volume of the ball in half. This implies that the second variation of volume is nonnegative for variations that have zero average. Therefore, $S(\varphi, \varphi) \geq 0$, a contradiction. $\square$

We remark that in the case of closed minimal hypersurfaces Σ^{n-1} in $\mathbb{S}^n$, the totally geodesic $\mathbb{S}^{n-1} \subset \mathbb{S}^n$ have index 1, and if Σ is not totally geodesic then index$(\Sigma) \geq n + 2$. In fact any minimal hypersurface in $\mathbb{S}^n$ that is not an equator comes equipped with $n + 1$ explicit eigenfunctions of the Jacobi operator with eigenvalue $-(n - 1)$, the components $N_1, \ldots, N_{n+1}$ of the unit normal N of Σ in $\mathbb{S}^n \subset \mathbb{R}^{n+1}$. Specifically, if $a \in \mathbb{R}^{n+1}$ then $u = \langle N, a \rangle$ satisfies $\Delta_\Sigma u + |A|^2 u = 0$ (the Jacobi equation of the minimal cone $C(\Sigma) \subset \mathbb{R}^{n+1}$), and the Jacobi operator associated with the second variation of Σ as a minimal hypersurface in $\mathbb{S}^n$ is $L = \Delta_\Sigma + |A|^2 + (n - 1)$, where here A denotes the second fundamental form of Σ in $\mathbb{S}^n$.

In contrast, for free boundary minimal hypersurfaces Σ^{n-1} in $\mathbb{B}^n$ no explicit eigenfunctions of the Jacobi operator with Robin boundary condition (1.7) are known. Although the components of the unit normal N of Σ in $\mathbb{B}^n$ satisfy the Jacobi equation $LN_i = \Delta_\Sigma N_i + |A|^2 N_i = 0$ on Σ, they do not satisfy the Robin boundary condition in general. On the other hand, even though we do not know explicit solutions of (1.7) in general, there are some interesting and important collections of functions for which the second variation of volume is negative, as we will discuss below.

First observe that if Σ is not an equatorial hyperplane and φ is a Steklov eigenfunction of Σ with eigenvalue $\sigma \leq 1$,

$$\begin{cases} \Delta\varphi = 0 & \text{on } \Sigma \\ \frac{\partial\varphi}{\partial\eta} = \sigma\varphi & \text{on } \partial\Sigma, \end{cases}$$

then

$$S(\varphi, \varphi) = -\int_\Sigma (\Delta\varphi + |A|^2\varphi)\,\varphi\,d\mu_\Sigma + \int_{\partial\Sigma}\left(\frac{\partial\varphi}{\partial\eta} - \varphi\right)\varphi\,d\mu_{\partial\Sigma}$$

$$= -\int_\Sigma |A|^2\,\varphi^2\,d\mu_\Sigma + (\sigma - 1)\int_{\partial\Sigma}\varphi^2\,d\mu_{\partial\Sigma}$$

$$< 0.$$

Recall that if Σ is a free boundary minimal submanifold in $\mathbb{B}^n$ then the coordinate functions $x_1, \ldots, x_n$ are Steklov eigenfunctions with eigenvalue 1 (see Sect. 1.2). Using these observations, we have the following.

- If Σ^{n-1} is a free boundary minimal hypersurface in $\mathbb{B}^n$ that is not an equatorial hyperplane, then index$(\Sigma) \geq n + 1$

Proof Since Σ is a free boundary minimal surface in $\mathbb{B}^n$, the coordinate functions $x_1, \ldots, x_n$ are Steklov eigenfunctions with eigenvalue 1, and the constant functions are Steklov eigenfunctions with eigenvalue 0. Let $\mathcal{V} = \text{span}\{1, x_1, \ldots, x_n\}$. First we claim that the index form S is negative definite on $\mathcal{V}$. Let $\varphi \in \mathcal{V}$. Then $\varphi = a + x \cdot v$ for some $a \in \mathbb{R}$ and $v \in \mathbb{R}^n$, and

$$S(\varphi, \varphi) = -\int_\Sigma (\Delta\varphi + |A|^2\varphi)\,\varphi\,d\mu_\Sigma + \int_{\partial\Sigma}\left(\frac{\partial\varphi}{\partial\eta} - \varphi\right)\varphi\,d\mu_{\partial\Sigma}$$

$$= -\int_\Sigma |A|^2\,\varphi^2\,d\mu_\Sigma + \int_{\partial\Sigma} -a(a + x \cdot v)\,d\mu_{\partial\Sigma}$$

$$= -\int_\Sigma |A|^2\varphi^2\,d\mu_\Sigma - \int_{\partial\Sigma} a^2\,d\mu_{\partial\Sigma}$$

$$< 0$$

if $\varphi \neq 0$, since Σ is not totally geodesic. Here in the last equality we have used that the coordinate functions are L^2-orthogonal to the constant functions on $\partial\Sigma$, since they are Steklov eigenfunctions corresponding to distinct eigenvalues,

$$0 = \int_\Sigma \Delta(x \cdot v)\,d\mu_\Sigma = \int_{\partial\Sigma} \frac{\partial x \cdot v}{\partial\eta}\,d\mu_{\partial\Sigma} = \int_\Sigma x \cdot v\,d\mu_{\partial\Sigma}.$$

Next we show that dim $\mathcal{V} = n + 1$. If dim $\mathcal{V} < n + 1$ then there exists $(a, v) \in \mathbb{R}^{n+1} \setminus \{0\}$ such that $a + x \cdot v = 0$ on Σ. But this means that Σ is contained in the hyperplane $x \cdot v = -a$, so $a = 0$ and Σ is an equatorial hyperplane, a contradiction. Therefore, index$(\Sigma) \geq n + 1$. $\qquad\square$

In particular, any free boundary minimal surface in $\mathbb{B}^3$ that is not an equatorial plane disk has index at least 4. Devyver [11], Smith and Zhou [46], and Tran [50] independently proved:

- The critical catenoid has index 4

Just as the Clifford torus is characterized as the unique closed minimal surface in $\mathbb{S}^3$ of index 5 [51], it is conjectured that the critical catenoid is the unique free boundary minimal surface in $\mathbb{B}^3$ of index 4:

Conjecture 1.5.3 If Σ free boundary minimal surface in $\mathbb{B}^3$ of index 4 then Σ is the critical catenoid.

In higher dimensions, the Clifford hypersurfaces in $\mathbb{S}^n$ have index $n + 2$ and are conjectured to be the only closed minimal hypersurfaces in $\mathbb{S}^n$ of index $n + 2$. In contrast, Smith et al. [47] recently proved that the higher dimensional free boundary minimal catenoids Σ^{n-1} in $\mathbb{B}^n$ have surprisingly high index. In general, we can say the following:

- If Σ^{n-1} is a free boundary minimal hypersurface in $\mathbb{B}^n$ of index $n + 1$, then the first Steklov eigenvalue $\sigma_1(\Sigma) = 1$

Proof Since Σ is a free boundary minimal hypersurface in $\mathbb{B}^n$, the coordinate functions are Steklov eigenfunctions with eigenvalue 1. Suppose $\sigma_1(\Sigma) \neq 1$. Then $\sigma_1(\Sigma) < 1$. Let u be a first Steklov eigenfunction, and consider $\mathcal{W} = \text{span}\{1, u, x_1, \ldots, x_n\}$. Since u is L^2-orthogonal to $\mathcal{V} = \text{span}\{1, x_1, \ldots, x_n\}$ on $\partial\Sigma$, we have that $\dim \mathcal{W} = 5$. By an argument similar to the one discussed above for $\mathcal{V}$, the index form S is negative definite on $\mathcal{W}$, and so Σ has index at least $n+2$, a contradiction. Therefore, $\sigma_1(\Sigma) = 1$. ☐

- If Σ^2 is a free boundary minimal *annulus* in $\mathbb{B}^3$ of index 4, then Σ is the critical catenoid.

Proof Since Σ has index 4, we have $\sigma_1(\Sigma) = 1$. That is, Σ is a free boundary minimal annulus such that the coordinate functions are *first* Steklov eigenfunctions. By Theorem 1.5.3, Σ is the critical catenoid. ☐

Therefore, to prove Conjecture 1.5.3 it suffices to prove that any free boundary minimal surface in $\mathbb{B}^3$ of index 4 is homeomorphic to an annulus.

In general, it is known that the index of a free boundary minimal surface in $\mathbb{B}^3$ tends to infinity as the genus or number of boundary components of Σ tends to infinity. This follows from index estimates of Sargent [44] and Ambrozio et al. [2]:

- If Σ is a free boundary minimal surface in $\mathbb{B}^3$ of genus γ with k boundary components, then

$$\text{index}(\Sigma) \geq \frac{1}{3}(2\gamma + k - 1).$$

The index estimates of [44] and [2] provide free boundary analogs of Savo's [45] index estimates for closed minimal hypersurfaces in $\mathbb{S}^n$.

As discussed above, if Σ is a free boundary minimal hypersurface in $\mathbb{B}^n$ that is not an equatorial hyperplane then any Steklov eigenfunction φ of Σ with eigenvalue ≤ 1 provides a variation of Σ with negative second variation $S(\varphi, \varphi) < 0$. There is another eigenvalue problem which is perhaps even more natural to consider when studying the index of free boundary minimal hypersurfaces in $\mathbb{B}^n$, and that is the Steklov problem associated with the Jacobi operator of Σ (rather than with the Laplacian of Σ). We will finish this section by describing work of Tran [50] which gives an interesting index characterization of free boundary minimal hypersurfaces in $\mathbb{B}^n$ in terms of the number of eigenvalues of the Dirichlet-to-Neumann operator associated with the Jacobi operator that are less than one.

Let $L = \Delta + |A|^2$ be the Jacobi operator of a free boundary minimal hypersurface Σ^{n-1} in $\mathbb{R}^n$. Observe that if u is a solution of

$$\begin{cases} Lu = 0 & \text{on } \Sigma \\ \frac{\partial u}{\partial \eta} = \sigma u & \text{on } \partial\Sigma \end{cases} \tag{1.8}$$

with $\sigma < 1$, then

$$S(u, u) = -\int_\Sigma u\, Lu + \int_{\partial\Sigma} \left(\frac{\partial u}{\partial \eta} - u\right) u = (\sigma - 1) \int_{\partial\Sigma} u^2 < 0.$$

We refer to (1.8) as the Steklov problem for the Jacobi operator L. Solutions of (1.8) are eigenfunctions of the Dirichlet-to-Neumann operator associated with the Jacobi operator (see [50]), which we now discuss.

In order to define the Dirichlet-to-Neumann operator associated with the Jacobi operator L, a first issue to consider is that there is an obstruction to the solvability of the Dirichlet problem for the Jacobi operator,

$$\begin{cases} L\hat{u} = 0 & \text{on } \Sigma \\ \hat{u} = u & \text{on } \partial\Sigma. \end{cases} \tag{1.9}$$

If $\hat{u}$ was a solution of (1.9) and $w \in \mathcal{J}_0^0 = \{w \in C^\infty(\Sigma) : Lw = 0 \text{ on } \Sigma, w = 0 \text{ on } \partial\Sigma\}$ was a solution of the Jacobi equation that vanished on the boundary, then

$$0 = \int_\Sigma (L\hat{u})\, w$$

$$= \int_\Sigma \hat{u}\, (Lw) + \int_{\partial\Sigma} \left(\frac{\partial \hat{u}}{\partial \eta} w - u \frac{\partial w}{\partial \eta}\right)$$

$$= \int_{\partial\Sigma} u \frac{\partial w}{\partial \eta}.$$

Conversely, by the Fredholm alternative, the Dirichlet problem (1.9) for L is solvable if and only if

$$\int_{\partial \Sigma} u \frac{\partial w}{\partial \eta} = 0 \qquad \text{for all } w \in \mathcal{J}_0^0,$$

and the solution is unique up to addition of $w \in \mathcal{J}_0^0$ ([50, Lemma 2.5], [11, Lemma 4.1]). That is, (1.9) is solvable if and only if $u \in (D_\eta \mathcal{J}_0^0)^\perp$, where

$$(D_\eta \mathcal{J}_0^0)^\perp := \{u \in C^\infty(\Sigma) : \int_{\partial \Sigma} u \frac{\partial w}{\partial \eta} = 0 \text{ for all } w \in \mathcal{J}_0^0\}.$$

Moreover, given $u \in (D_\eta \mathcal{J}_0^0)^\perp$, there is a *unique* solution $\hat{u}$ of (1.9) such that $\frac{\partial \hat{u}}{\partial \eta} \in (D_\eta \mathcal{J}_0^0)^\perp$. To see that such a solution $\hat{u}$ exists, let $\hat{u}_0$ be any solution of (1.9) and let Π denote the L^2-orthogonal projection of $L^2(\partial \Sigma)$ onto $D_\eta \mathcal{J}_0^0$. Then $\Pi(\frac{\partial \hat{u}_0}{\partial \eta}) = \frac{\partial v}{\partial \eta}$ for some $v \in \mathcal{J}_0^0$, and $\hat{u} = \hat{u}_0 - v$ is the desired solution of (1.9). Clearly if such a solution exists, then it is unique, since if $\hat{u}_1$ and $\hat{u}_2$ were two such extensions, then

$$\int_{\partial \Sigma} \frac{\partial}{\partial \eta} (\hat{u}_1 - \hat{u}_2) \frac{\partial w}{\partial \eta} = 0 \qquad \text{for all } w \in \mathcal{J}_0^0,$$

and so in particular,

$$\int_{\partial \Sigma} \frac{\partial}{\partial \eta} (\hat{u}_1 - \hat{u}_2) \frac{\partial}{\partial \eta} (\hat{u}_1 - \hat{u}_2) = 0,$$

which implies that $\frac{\partial(\hat{u}_1 - \hat{u}_2)}{\partial \eta} = 0$, and so $\hat{u}_1 - \hat{u}_2 = 0$ since $\hat{u}_1 - \hat{u}_2 \in \mathcal{J}_0^0$.

Therefore, there exists a well-defined Dirichlet-to-Neumann operator T associated with the Jacobi operator L,

$$T : (D_\eta \mathcal{J}_0^0)^\perp \to (D_\eta \mathcal{J}_0^0)^\perp$$

given by

$$Tu = \frac{\partial \hat{u}}{\partial \eta}$$

where $\hat{u}$ is the uniquely defined solution of (1.9) as above. T is self-adjoint with discrete spectrum [50]. Let E_σ denote the eigenspace corresponding to an eigenvalue σ of T.

As observed above, if $u \in E_\sigma$ is an eigenfunction of T with eigenvalue $\sigma < 1$, then the second variation of volume of Σ is negative $S(u, u) < 0$. The following is an interesting characterization of the index in terms of the number of eigenvalues of the Dirichlet-to-Neumann operator associated with the Jacobi operator that are less than one, due to H. Tran [50]:

Theorem 1.5.6 ([50]) *Suppose Σ^{n-1} is a free boundary minimal hypersurface in $\mathbb{B}^n$. Then*

$$index(\Sigma) = \dim \bigoplus_{\sigma < 1} E_\sigma + \dim \bigoplus_{\lambda \leq 0} V_\lambda$$

where V_λ denotes the space of eigenfunctions of the Jacobi operator L with eigenvalue λ with Dirichlet boundary condition:

$$\begin{cases} Lu + \lambda u = 0 & \text{on } \Sigma \\ u = 0 & \text{on } \partial\Sigma. \end{cases}$$

Acknowledgments The author is grateful to M. Gursky and A. Malchiodi for the invitation to and for organizing the CIME summer school on Geometric Analysis in 2018, for which these lecture notes were written. Much of the content of these notes is based on joint papers with R. Schoen and the author is grateful to R. Schoen for his support and inspiration.

A. Fraser was partially supported by the Natural Sciences and Engineering Research Council of Canada and a Simons Visiting Professorship at MFO and CIME. Part of this article was written while the author was visiting the Institute for Advanced Study for the special year *Variational Methods in Geometry*, with funding from NSF grant DMS-1638352 and the James D. Wolfensohn Fund.

References

1. F.J. Almgren, Some interior regularity theorems for minimal surfaces and an extension of Bernstein's theorem. Ann. Math. (2) **84**, 277–292 (1966)
2. L. Ambrozio, A. Carlotto, B. Sharp, Index estimates for free boundary minimal hypersurfaces. Math. Ann. **370**(3–4), 1063–1078 (2018)
3. S. Brendle, Embedded minimal tori in S^3 and the Lawson conjecture. Acta Math. **211**(2), 177–190 (2013)
4. F. Brock, An isoperimetric inequality for eigenvalues of the Stekloff problem. ZAAM Z. Angew. Math. Mech. **81**, 69–71 (2001)
5. D. Bucur, V. Ferone, C. Nitsch, C. Trombetti, Weinstock inequality in higher dimensions. J. Differ. Geom. To appear
6. E. Calabi, Minimal immersions of surfaces in Euclidean spheres. J. Differ. Geom. **1**, 111–125 (1967)
7. A. Carlotto, G. Franz, M. Schulz, Free boundary minimal surfaces with connected boundary and arbitrary genus. arXiv:2001.04920
8. D. Cianci, A. Girouard, Large spectral gaps for Steklov eigenvalues under volume constraints and under localized conformal deformations. Ann. Global Anal. Geom. **54**(4), 529–539 (2018)
9. B. Colbois, A. El Soufi, A. Girouard, Isoperimetric control of the Steklov spectrum. J. Funct. Anal. **261**(5), 1384–1399 (2011)
10. B. Colbois, A. Girouard, B. Raveendran, The Steklov spectrum and coarse discretizations of manifolds with boundary. Pure Appl. Math. Q. **14**(2), 357–392 (2018)
11. B. Devyver, Index of the critical catenoid. Geom. Dedicata. **199**, 355–371 (2019)

12. B. Dittmar, Sums of reciprocal Stekloff eigenvalues. Math. Nachr. **268**, 44–49 (2004)
13. A. El Soufi, S. Ilias, Immersions minimales, première valeur propre du laplacien et volume conforme. Math. Ann. **275**(2), 257–267 (1986)
14. A. El Soufi, H. Giacomini, M. Jazar, A unique extremal metric for the least eigenvalue of the Laplacian on the Klein bottle. Duke Math. J. **135**, 181–202 (2006)
15. A. Folha, F. Pacard, T. Zolotareva, Free boundary minimal surfaces in the unit 3-ball. Manuscripta Math. **154**(3–4), 359–409 (2017)
16. A. Fraser, M. Li, Compactness of the space of embedded minimal surfaces with free boundary in three-manifolds with nonnegative Ricci curvature and convex boundary. J. Differ. Geom. **96**(2), 183–200 (2014)
17. A. Fraser, R. Schoen, The first Steklov eigenvalue, conformal geometry, and minimal surfaces. Adv. Math. **226**(5), 4011–4030 (2011)
18. A. Fraser, R. Schoen, Minimal surfaces and eigenvalue problems, in *Geometric Analysis, Mathematical Relativity, and Nonlinear Partial Differential Equations*. Contemporary Mathematics, vol. 599 (American Mathematical Society, Providence, 2013), pp. 105–121
19. A. Fraser, R. Schoen, Uniqueness theorems for free boundary minimal disks in space forms. Int. Math. Res. Not. **2015**(17), 8268–8274 (2015)
20. A. Fraser, R. Schoen, Sharp eigenvalue bounds and minimal surfaces in the ball. Invent. Math. **203**(3), 823–890 (2016)
21. A. Fraser, R. Schoen, Shape optimization for the Steklov problem in higher dimensions. Adv. Math. **348**, 146–162 (2019)
22. A. Fraser, R. Schoen, Some results on higher eigenvalue optimization (2019). arXiv:1910.03547
23. A. Girouard, I. Polterovich, On the Hersch-Payne-Schiffer estimates for the eigenvalues of the Steklov problem. Funct. Anal. Appl. **44**(2), 106–117 (2010)
24. A. Girouard, I. Polterovich, Spectral geometry of the Steklov problem. J. Spectr. Theory **7**(2), 321–359 (2017)
25. J. Hersch, Quatre propriétés isopérimétriqes de membranes sphériques homogènes. C.R. Acad. Sci. Paris Sér. A-B **270**, A1645–A1648 (1970)
26. D. Jakobson, M. Levitin, N. Nadirashvili, N. Nigam, I. Polterovich, How large can the first eigenvalue be on a surface of genus two? Int. Math. Res. Not. **63**(63), 3967–3985 (2005)
27. D. Jakobson, N. Nadirashvili, I. Polterovich, Extremal metric for the first eigenvalue on a Klein bottle. Canad. J. Math. **58**, 381–400 (2006)
28. N. Kapouleas, M. Li, Free boundary minimal surfaces in the unit three-ball via desingularization of the critical catenoid and the equatorial disk. arXiv:1709.08556
29. N. Kapouleas, D. Wiygul, Free-boundary minimal surfaces with connected boundary in the 3-ball by tripling the equatorial disc. arXiv:1711.00818
30. M. Karpukhin, Upper bounds for the first eigenvalue of the Laplacian on non-orientable surfaces. Int. Math. Res. Not. IMRN **2016**(20), 6200–6209 (2016)
31. M. Karpukhin, Maximal metrics for the first Steklov eigenvalue on surfaces (2018). arXiv:1801.06914
32. M. Karpukhin, On the Yang-Yau inequality for the first Laplace eigenvalue. Geom. Funct. Anal. **29**(6), 1864–1885 (2019)
33. D. Ketover, Free boundary minimal surfaces of unbounded genus (2016). arXiv:1612.08691
34. G. Kokarev, Variational aspects of Laplace eigenvalues on Riemannian surfaces. Adv. Math. **258**, 191–239 (2014)
35. H.B. Lawson, Jr., Complete minimal surfaces in S^3. Ann. Math. (2) **92**, 335–374 (1970)
36. M. Li, Free boundary minimal surfaces in the unit ball: recent advances and open questions, in *Proceedings of the First Annual Meeting of the ICCM* (2019)
37. P. Li, S.-T. Yau, A new conformal invariant and its applications to the Willmore conjecture and the first eigenvalue of compact surfaces. Invent. Math. **69**(2), 269–291 (1982)
38. H. Matthiesen, R. Petrides, Monotonicity results for the first Steklov eigenvalue on compact surfaces. arXiv:1801.08518

39. P. McGrath, A characterization of the critical catenoid. Indiana Univ. Math. J. **67**(2), 889–897 (2018)
40. S. Montiel, A. Ros, Minimal immersions of surfaces by the first eigenfunctions and conformal area. Invent. Math. **83**(1), 153–166 (1985)
41. N. Nadirashvili, Berger's isoperimetric problem and minimal immersions of surfaces. Geom. Funct. Anal. **6**(5), 877–897 (1996)
42. S. Nayatani, T. Shoda, Metrics on a closed surface of genus two which maximize the first eigenvalue of the Laplacian. C. R. Math. Acad. Sci. Paris **357**(1), 84–98 (2019)
43. J.C.C. Nitsche, Stationary partitioning of convex bodies. Arch. Ration. Mech. Anal. **89**(1), 1–19 (1985)
44. P. Sargent, Index bounds for free boundary minimal surfaces of convex bodies. Proc. Amer. Math. Soc. **145**(6), 2467–2480 (2017)
45. A. Savo, Index bounds for minimal hypersurfaces of the sphere. Indiana Univ. Math. J. **59**(3), 823–837 (2010)
46. G. Smith, D. Zhou, The Morse index of the critical catenoid. Geom. Dedicata **201**, 13–19 (2019)
47. G. Smith, A. Stern, H. Tran, D. Zhou, On the Morse index of higher-dimensional free boundary minimal catenoids (2017). arXiv:1709.00977
48. R. Souam, On stability of stationary hypersurfaces for the partitioning problem for balls in space forms. Math. Z. **224**(2), 195–208 (1997)
49. G. Szegö, Inequalities for certain eigenvalues of a membrane of given area. J. Ration. Mech. Anal. **3**, 343–356 (1954)
50. H. Tran, Index characterization for free boundary minimal surfaces. Comm. Anal. Geom. **28**(1), 189–222 (2020)
51. F. Urbano, Minimal surfaces with low index in the three-dimensional sphere. Proc. Amer. Math. Soc. **108**(4), 989–992 (1990)
52. H.F. Weinberger, An isoperimetric inequality for the N-dimensional free membrane problem. J. Ration. Mech. Anal. **5**, 633–636 (1956)
53. R. Weinstock, Inequalities for a classical eigenvalue problem. J. Ration. Mech. Anal. **3**, 745–753 (1954)
54. P. Yang, S.-T. Yau, Eigenvalues of the Laplacian of compact Riemann surfaces and minimal submanifolds. Ann. Scuola Norm. Sup. Pisa Cl. Sci. (4) **7**(1), 55–63 (1980)

Chapter 2
Applications of Min–Max Methods to Geometry

Fernando C. Marques and André Neves

Abstract The existence of minimal surfaces in closed manifolds is a classical subject with a long history. This chapter presents some recent advances on the subject, motivated by *Yau's conjecture* concerning the existence of infinitely-many ones. The main tools used here are a combination of techniques from Geometric Measure Theory and Minimal methods. The conjecture is proved for a large class of metrics and, via the concept of *volume spectrum*, a density result is also derived.

2.1 Introduction

Let (M^{n+1}, g) be an $(n+1)$-dimensional closed Riemannian manifold. We assume, for convenience, that (M, g) is isometrically embedded in some Euclidean space $\mathbb{R}^J$.

A closed embedded hypersurface $\Sigma \subset M$ is called a *minimal hypersurface* if it is a critical point for the area functional, meaning that for every ambient vector field X in M we have

$$\frac{d}{dt}\text{vol}(\phi_t(\Sigma))_{t=0} = 0,$$

The first author is partly supported by NSF-DMS-1811840. The second author is partly supported by NSF DMS-1710846 and by a Simons Investigator Grant.

F. C. Marques
Institute for Advanced Study, Princeton University Princeton, Princeton, NJ, USA
e-mail: coda@ias.edu; coda@math.princeton.edu

A. Neves (✉)
University of Chicago, Department of Mathematics, Chicago, IL, USA
e-mail: aneves@uchicago.edu

M. J. Gursky, A. Malchiodi (eds.), *Geometric Analysis*, Lecture Notes
in Mathematics 2263, https://doi.org/10.1007/978-3-030-53725-8_2

where $\{\phi_t\}_{t\in\mathbb{R}}$ is a one-parameter family of diffemorphisms generated by the vector field X. From the first variation formula we know that

$$\frac{d}{dt}\text{vol}(\phi_t(\Sigma))_{t=0} = -\int_\Sigma \langle H, X\rangle d\Sigma,$$

where H is the mean curvature vector of Σ, and so minimal hypersurfaces are those which have $H = 0$.

The simplest example of a minimal surface in $\mathbb{R}^3$ is given by plane and in the unit 3-sphere $S^3 \subset \mathbb{R}^4$ simple examples can be given by equators (intersection of a hyperplane in $\mathbb{R}^4$ with S^3). Many more examples exist in $\mathbb{R}^3$ (like the catenoid, helicoid, or Schwarz P surface) and Lawson [15] showed in the 1970s that S^3 has closed orientable minimal surfaces of arbitrary genus.

One of the most fundamental questions one can ask regarding closed minimal hypersurfaces is whether they exist and this was answered in the early 1980s through the combined work of Almgren [2], Pitts [24] and Schoen and Simon.

Theorem 2.1.1 *Every closed Riemannian manifold (M^{n+1}, g) has a closed minimal hypersurface that is smooth and embedded outside a set of Hausdorff dimension less than or equal to $n - 7$.*

Around the same time Yau [32] made the following conjecture:

Yau's Conjecture 2.1.2 *Every closed Riemannian three-manifold contains infinitely many smooth, closed minimal surfaces.*

The purpose of these notes is to present some of the recent progress made regarding Yau's Conjecture and the existence of minimal hypersurface. For the sake of brevity, we will not do an exhaustive account of the historical developments (which means we will not mention the long list of beautiful results regarding existence of geodesics), nor will we cover all the recent developments in neighboring areas (such as free boundary minimal surfaces or the Allen–Cahn regularization). We focus mainly in providing the background needed in order to prove some of the recent developments.

Around the time the conjecture was made, the combined work of Almgren [2], Pitts [24] Schoen and Simon [25] showed the following result:

Theorem 2.1.3 *Every closed Riemannian manifold (M^{n+1}, g) has a closed minimal hypersurface that is smooth and embedded outside a set of Hausdorff dimension less than or equal to $n - 7$.*

Not much progress was done regarding Yau's conjecture until we showed [18] (see also [18, Remark 1.6]) the following result:

Theorem 2.1.4 *Every closed Riemannian manifold (M^{n+1}, g) with positive Ricci curvature has infinitely many distinct minimal hypersurfaces that are smooth and embedded outside a set of Hausdorff dimension less than or equal to $n - 7$.*

Recently, jointly with Irie [13], we showed a denseness result that implies Yau's conjecture in the generic case.

Denseness Theorem 2.1.5 *Let M^{n+1} be a closed manifold of dimension $(n + 1)$, with $3 \leq (n + 1) \leq 7$.*

For a C^∞-generic Riemannian metric g on M, the union of all closed, smooth, embedded minimal hypersurfaces is dense.

Later, jointly with Song [22], we showed the existence of a sequence of closed embedded minimal hypersurfaces that becomes equidistributed.

Equidistribution Theorem 2.1.6 *Let M^{n+1} be a closed manifold of dimension $(n + 1)$, with $3 \leq (n + 1) \leq 7$.*

For a C^∞-generic Riemannian metric g on M, there exists a sequence $\{\Sigma_j\}_{j\in\mathbb{N}}$ of closed, smooth, embedded, connected minimal hypersurfaces that is equidistributed in M: for any $f \in C^0(M)$ we have

$$\lim_{q \to \infty} \frac{1}{\sum_{j=1}^{q} vol_g(\Sigma_j)} \sum_{j=1}^{q} \int_{\Sigma_j} f \, d\Sigma_j = \frac{1}{vol_g M} \int_M f dV.$$

Actually, the equidistribution proven in [22] is slightly more general because the test functions are allowed to be symmetric 2-tensors.

Shortly after these results were proven, two serious contributions to the field were made: Firstly, Song [28] settled Yau's conjecture by showing the following result.

Theorem 2.1.7 *Every closed Riemannian manifold (M^{n+1}, g) with $3 \leq (n+1) \leq 7$ has infinitely many distinct closed, smooth, embedded minimal hypersurfaces.*

Secondly, X. Zhou [35] used a novel regularization of the area functional (developed by him and Zhu in [34, 36]) to prove the Multiplicity One Conjecture proposed by the authors in [21] (see also [20]).

Theorem 2.1.8 *Let (M^{n+1}, g) be a closed Riemannian manifold, $3 \leq (n + 1) \leq 7$, with a bumpy metric.*

If Π is a non-trivial homotopy class there is an embedded, two-sided, multiplicity one, minimal hypersurface Σ with

$$\mathbf{L}(\Pi) = \text{vol}(\Sigma).$$

Before we describe a consequence of his work we need to introduce some more concepts.

Consider Σ a closed minimal hypersurface of M and let $N\Sigma$, $\Gamma(N\Sigma)$ denote, respectively, the normal bundle of Σ and the space of sections of $N\Sigma$. The second variation of Σ is a quadratic form on $\Gamma(N\Sigma)$ defined by

$$\delta^2\Sigma(X, X) := \frac{d^2}{dt^2}\text{vol}(\phi_t(\Sigma))|_{t=0}$$

$$= \int_\Sigma |\nabla^\perp X|^2 - \text{Ric}(X, X) - |A|^2|X|^2 d\Sigma,$$

where $X \in \Gamma(N\Sigma)$, $\{\phi_t\}_{t\in\mathbb{R}}$ denotes the one-parameter family of diffemorphisms generated by X (after being extended to vector field on M), $\nabla^\perp$ is the natural connection on $N\Sigma$, and $|A|^2$ is the norm of the second fundamental form. Elements in the kernel of $\delta^2\Sigma$ are called *Jacobi vector fields*.

White [30] (see also [31]) proved a Bumpy Metrics Theorem which says that almost every metric (in the Baire category sense) is bumpy, i.e., every minimal hypersurface has no non-trivial Jacobi vector fields.

The Morse index of Σ is the largest possible dimension of a vector subspace $P \subset \Gamma(N\Sigma)$ so that $\delta^2\Sigma$ is a negative quadratic form when restricted to P. Intuitively speaking, the Morse index of Σ (denoted by index(Σ)) is the number of linearly independent deformations that strictly decrease the volume of Σ. For instance, on the unit 3-sphere $S^3 \subset \mathbb{R}^4$, the Morse index of an equator (intersection of a hyperplane in $\mathbb{R}^4$ with S^3) is one because normal deformations decrease the area and volume preserving deformations never decrease the area. One can find an ellipsoid in $\mathbb{R}^4$ so that the intersection of the ellipsoid with each hyperplane $\{x_i = 0\}$ is a minimal sphere with Morse index i, $i = 1, \dots, 4$. On $\mathbb{RP}^3$ with the round metric, the Morse index of a equatorial $\mathbb{RP}^2$ is zero because it is area-minimizing in its homotopy class.

Combining Zhou's solution to the Multiplicity One Conjecture with the Morse index bounds proven by the authors in [21] and with the Weyl Law for the Volume Spectrum proven by Liokumovich and the authors in [16] we have

Theorem 2.1.9 *Assume (M^{n+1}, g) is a closed Riemannian manifold, $3 \le (n+1) \le 7$, with a bumpy metric.*

For each $k \in \mathbb{N}$ there is an embedded, two-sided, multiplicity one, minimal hypersurface Σ_k with

$$\text{vol}(\Sigma_k) \simeq a(n)\text{vol}(M)^{\frac{n}{n+1}} k^{\frac{1}{n+1}} \quad and \quad \text{index}(\Sigma_k) = k,$$

where $a(n)$ is a universal constant.

2.1.1 Organization

In Sect. 2.2 we introduce the basic concepts and describe the main results of Min–max Theory for minimal hypersurfaces. In particular, in Sect. 2.2.3 we explain how Theorem 2.1.1 follows from Min–max Theory. Section 2.3 is dedicated to the concept of volume spectrum and to the proof of the Weyl Law for the volume spectrum. In Sect. 2.3.3 we explain how Theorem 2.3.10 follows from the solution to the Multiplicity One Conjecture, lower bounds for Morse index, and Weyl Law for the Volume Spectrum. In Sect. 2.4.1 we prove the Denseness Theorem 2.1.5 and the proof of the Equidistribution Theorem 2.1.6 is sketched in Sect. 2.4.2.

2.2 Min–Max Theory

2.2.1 Basic Notions in Geometric Measure Theory

The following definitions are taken from [27] and they correspond to extensions of the concept of a submanifold. In a nutshell, we will be working mostly with the space of mod 2 codimension one cycles $\mathcal{Z}_k(M; \mathbb{Z}_2)$ which can thought of as the space of all closed hypersurfaces in M. The reader comfortable with these concepts can skip this section.

A set $S \subset \mathbb{R}^J$ is *countable k-rectifiable* if $S \subset S_0 \cup_{j \in \mathbb{N}} S_j$, where $\mathcal{H}^k(S_0) = 0$ and S_j, $j \in \mathbb{N}$, is an embedded k-dimensional C^1-submanifold. We assume in addition that the set $S \subset \mathbb{R}^J$ is $\mathcal{H}^k$-measurable and $\mathcal{H}^k(S \cap K) < +\infty$ for every compact set $K \subset \mathbb{R}^J$. In this case, k-rectifiable sets are characterized by the property that they have a well defined k-dimensional tangent plane $T_x S$ for $\mathcal{H}^k$-a.e. $x \in S$ (see [27, Theorem 11.6]). Let S_* denote the subset of S for which the k-dimensional tangent plane is well defined.

The Grassmanian of k-planes in $\mathbb{R}^J$ is denoted by $G_k(\mathbb{R}^J)$. There is a natural projection π from $G_k(\mathbb{R}^J)$ onto $\mathbb{R}^J$. A *rectifiable k-varifold* V is a Radon measure on $G_k(\mathbb{R}^J)$ so that for every measurable set $A \subset G_k(\mathbb{R}^J)$

$$V(A) = \int_{S \cap \pi(TS \cap A)} \theta(x) d\mathcal{H}^k$$

where S is a countable k-rectifiable set, θ is a positive locally $\mathcal{H}^k$-integrable function on S, and $TS = \{(x, T_x S) : x \in S_*\}$. There is a natural Radon measure $||V||$ on $\mathbb{R}^J$ defined as $||V||(A) = V(\pi^{-1}(A))$ for every measurable set $A \subset \mathbb{R}^J$. We say that $||V||(\mathbb{R}^J)$ is the *mass* of V and is the analogue of k-dimensional volume.

We denote by $\mathcal{V}_n(M)$ the closure, in the weak topology, of the space of rectifiable k-varifolds in $\mathbb{R}^J$ with support contained in M. When the function θ is $\mathbb{N}$-valued, V is called an *integer k-varifold*.

Denote by $\mathcal{D}^k(\mathbb{R}^J)$ the set of smooth k-forms of $\mathbb{R}^J$ with compact support. Given an element $\omega \in \mathcal{D}^k(\mathbb{R}^J)$ we define $|\omega| = \sup_{x \in \mathbb{R}^J}\{\langle \omega(x), \omega(x) \rangle^{1/2}\}$.

A *k-current* T is a continuous linear functional on $\mathcal{D}^k(\mathbb{R}^J)$. Its boundary ∂T is a $k - 1$-current that is defined as $\partial T(\phi) = T(d\phi)$, $\phi \in \mathcal{D}^{k-1}(\mathbb{R}^J)$. Naturally, $\partial^2 T = 0$. We will assume that every k-current has compact support. The restriction of a current T to an open set U is denoted by $T \llcorner U$.

Following [27, Section 27], we say that T is an *integer multiplicity k-current* (or simply integer multiplicity current) if it can be expressed as

$$T(\phi) = \int_S \langle \phi(x), \tau(x) \rangle \theta(x) d\mathcal{H}^k, \quad \phi \in \mathcal{D}^k(\mathbb{R}^J),$$

where S is a $\mathcal{H}^k$-measurable countable k-rectifiable set, θ is a $\mathcal{H}^k$-integrable $\mathbb{N}$-valued function, and τ is a k-form so that for all $x \in S_*$, $\tau(x)$ is a volume form for $T_x S$. In particular, $\tau(x)$ chooses an orientation for $T_x S$. The *mass* of an integer multiplicity k-current T is defined as

$$\mathbf{M}(T) = \sup\{T(\phi) : \phi \in \mathcal{D}^k(\mathbb{R}^L), |\phi| \leq 1\}$$

The space of integral currents with finite mass corresponds to the space of rectifiable currents defined in [7, 4.1.24] (see [7, Theorem 4.1.28]).

The space of k-currents T such that both T and ∂T are integer multiplicity currents with finite mass and support contained in M is denoted by $\mathbf{I}_k(M)$. This space is called the space of *integral k-currents*. The space of k-cycles is defined as those elements $T \in \mathbf{I}_k(M)$ so that $T = \partial Q$ for some $Q \in \mathbf{I}_{k+1}(M)$ and is denoted at $\mathcal{Z}_k(M)$. Note that in our notation $\mathcal{Z}_k(M)$ stands for the connected component containing zero of the set of integral currents with no boundary (thus differing slightly from the notation in [27] or [7]).

Given $T \in \mathbf{I}_k(M)$ there is a natural varifold $|T|$ associated to it and we denote its Radon measure by $\|T\|$. We have $\|T\|(M) = \mathbf{M}(T)$. The following varifolds appear naturally in the context of min–max theory.

Definition 2.2.1 We say an integer n-varifold V is a *smooth embedded minimal cycle* if there is a disjoint collection $\{\Sigma_1, \ldots, \Sigma_N\}$ of closed, smooth, embedded, minimal hypersurfaces in M and a set of integers $\{m_1, \ldots, m_N\} \subset \mathbb{N}$, such that

$$V = m_1|\Sigma_1| + \cdots + m_N|\Sigma_N|.$$

The spaces above come with several relevant metrics. Given $T_1, T_2 \in \mathbf{I}_k(M)$, the *flat metric* is defined by

$$\mathcal{F}(T_1, T_2) = \inf\{\mathbf{M}(Q) + \mathbf{M}(R) : T_1 - T_2 = Q + \partial R, P \in \mathbf{I}_k(M), Q \in \mathbf{I}_{k+1}(M)\}$$

and induces the so called *flat topology* on $\mathbf{I}_k(M)$. We also use $\mathcal{F}(T) = \mathcal{F}(T, 0)$ and one has that

$$\mathcal{F}(T) \leq \mathbf{M}(T) \quad \text{for all } T \in \mathbf{I}_k(M).$$

The **F**-*metric* on $\mathcal{V}_k(M)$ is defined in the book of Pitts [24, page 66] as:

$$\mathbf{F}(V, W) = \sup\{V(f) - W(f) : f \in C_c(G_k(\mathbb{R}^L)),$$
$$|f| \leq 1, \mathrm{Lip}(f) \leq 1\},$$

for $V, W \in \mathcal{V}_k(M)$ and induces the varifold weak topology on

$$\mathcal{V}_n(M) \cap \{V : ||V||(M) \leq c\}$$

for any $c > 0$.

Finally, the **F**-*metric* on $\mathbf{I}_k(M)$ is defined by

$$\mathbf{F}(S, T) = \mathcal{F}(S - T) + \mathbf{F}(|S|, |T|).$$

We have $\mathbf{F}(|S|, |T|) \leq \mathbf{M}(S - T)$ and hence $\mathbf{F}(S, T) \leq 2\mathbf{M}(S - T)$ for any $S, T \in \mathbf{I}_l(M)$.

We assume that $\mathbf{I}_k(M)$ and $\mathcal{Z}_k(M)$ have the topology induced by the flat metric. Informally, $T, S \in \mathcal{Z}_k(M)$ being very close to each other in the flat metric means that $T - S$ is the boundary of $Q \in \mathbf{I}_{k+1}(M)$ with very small mass. When endowed with the topology of the **F**-metric these spaces will be denoted by $\mathbf{I}_k(M; \mathbf{F})$ and $\mathcal{Z}_n(M; \mathbf{F})$, respectively.

The Federer–Fleming Compactness Theorem [7, 4.2.17] states that the set

$$\{T \in \mathcal{Z}_k(M) : \mathbf{M}(T) \leq C\}, \quad C > 0$$

is compact in the flat topology.

An important fact in the theory is that, while the mass is continuous in the varifold topology, it is only lower semicontinuous in the flat topology. The loss of mass in the limit is illustrated with the following standard example: let

$$Q_i = \{(x, y) \in \mathbb{R}^2 : 0 \leq x \leq 1, 0 \leq y \leq i^{-1}\}.$$

Then ∂Q_i tends to zero in the flat topology, but $\mathbf{M}(\partial Q_i)$ tends to 2. In this example $|\partial Q_i|$ tends to $2([0, 1] \times \{0\})$ in the varifold topology.

For our purposes, we are interested in the space of mod 2 integral k-currents or mod 2 k-cycles that we denote by $\mathbf{I}_k(M; \mathbb{Z}_2)$ and $\mathcal{Z}_k(M; \mathbb{Z}_2)$, respectively. This space is defined via an equivalence relation, where we say that $T \equiv S$ if $T - S = 2Q$, T, S, Q being in $\mathbf{I}_k(M)$, and they were first introduced by Ziemer [33]. All the concepts we mentioned for $\mathbf{I}_k(M)$ and $\mathcal{Z}_k(M)$ can be extended to $\mathbf{I}_k(M; \mathbb{Z}_2)$ and $\mathcal{Z}_k(M; \mathbb{Z}_2)$ as well (see [33] or [8]). The Constancy Theorem [27, Theorem 26.27] says that if $T \in \mathcal{I}_{n+1}(M; \mathbb{Z}_2)$ has $\partial T = 0$, then either $T = M$ or $T = 0$.

The Isoperimetric Inequality of Federer–Fleming (adapted to mod 2 integral currents in [33, Corollary 4.7]) gives constants a_M, b_M so that for every $T \in \mathcal{Z}_n(M; \mathbb{Z}_2)$ with $\mathbf{M}(T) \leq a_M$ there is Ω in $\mathbf{I}_{n+1}(M; \mathbb{Z}_2)$ such that

$$\partial \Omega = T \quad \text{and} \quad \mathbf{M}(\Omega) \leq b_M \mathbf{M}(T)^{\frac{n+1}{n}}. \tag{2.2.1}$$

When combined with the Constancy Theorem we obtain the following lemma:

Lemma 2.2.2 *There is ε so that for every $T \in \mathcal{Z}_n(M; \mathbb{Z}_2)$ with $\mathcal{F}(T) < \varepsilon$ there is a unique $S \in \mathbf{I}_{n+1}(M; \mathbb{Z}_2)$ with $\mathcal{F}(T) = \mathbf{M}(S)$.*

Proof Choose Q and R so that $T = Q + \partial R$ and $\mathbf{M}(Q) + \mathbf{M}(R) \leq \varepsilon$. Assuming $\varepsilon \leq \min\{a_M, b_M^{-n}, \mathrm{vol}(M)/3\}$ we have from the Isoperimetric Inequality the existence of Ω with $\partial \Omega = Q$ and $\mathbf{M}(\Omega) \leq \mathbf{M}(Q)$. As a result, setting $S = \Omega + R$, we have $T = \partial S$ and

$$\mathbf{M}(S) \leq \mathbf{M}(\Omega) + \mathbf{M}(R) \leq \mathbf{M}(Q) + \mathbf{M}(R) \leq \mathrm{vol}(M)/3.$$

From the Constancy Theorem we have that if $S' \in \mathbf{I}_{n+1}(M; \mathbb{Z}_2)$ is such that $\partial S' = T$ and $\mathbf{M}(S') \leq \mathrm{vol}(M)/3$ then $S = S'$. This implies the lemma. $\qquad\square$

2.2.2 Space of Cycles

The basic principle of min–max theory is to use the homotopy classes of $\mathcal{Z}_n(M; \mathbb{Z}_2)$ to produce minimal hypersurfaces and so it is important that we understand the topology of $\mathcal{Z}_n(M; \mathbb{Z}_2)$. There is a map from $\mathbb{RP}^\infty$ to $\mathcal{Z}_n(M; \mathbb{Z}_2)$ that we now describe.

Let $f : M \to \mathbb{R}$ be a Morse function, with $f(M) = [0, 1]$, and consider the map $\hat{\Phi} : \mathbb{RP}^\infty \to \mathcal{Z}_n(M; \mathbb{Z}_2)$ given by

$$\hat{\Phi}([a_0 : a_1 : \cdots : a_k : 0 : \cdots]) = \partial \{x \in M : a_0 + a_1 f(x) + \cdots + a_k f(x)^k \leq 0\}.$$

The map is well defined because we are considering mod 2 cycles. In [18] (Claim 5.6), we proved the map $\hat{\Phi}$ is continuous in the flat topology.

Theorem 2.2.3 *The map $\hat{\Phi}$ is a weak homotopy equivalence.*

Almgren computed in [1] the homotopy groups of $\mathcal{Z}_k(M; \mathbb{Z}_2)$ for all $0 \leq k \leq n+1$ but the proof is more complicated than the argument we present (see [21, Section 5]).

Proof Consider the continuous map

$$\partial : \mathbf{I}_{n+1}(M; \mathbb{Z}_2) \to \mathcal{Z}_n(M; \mathbb{Z}_2).$$

From the Constancy Theorem we know that $\partial U = \partial V$ implies that $U = V$ or $U = M - V$, which means that the map is 2 to 1.

Claim 1 $\mathbf{I}_{n+1}(M; \mathbb{Z}_2)$ is contractible.

We define $H : [0, 1] \times \mathbf{I}_{n+1}(M; \mathbb{Z}_2) \to \mathbf{I}_{n+1}(M; \mathbb{Z}_2)$ by

$$H(t, U) = U \llcorner \{f \leq t\}.$$

The map H is continuous, $H(1, U) = U$ and $H(0, U) = 0$ for every $U \in \mathbf{I}_{n+1}(M; \mathbb{Z}_2)$. This proves the claim.

From the definition of $\mathcal{Z}_n(M; \mathbb{Z}_2)$ we have that the map ∂ is surjective and so it follows from the previous claim that $\mathcal{Z}_n(M; \mathbb{Z}_2)$ is path-connected.

Claim 2 $\mathbf{I}_{n+1}(M; \mathbb{Z}_2)$ is a covering space.

We need to find an open cover $\{B_T\}_{T \in \mathcal{Z}_n(M;\mathbb{Z}_2)}$ of $\mathcal{Z}_n(M; \mathbb{Z}_2)$ such that each $\partial^{-1}(B_T)$ is a disjoint union of open sets in $\mathbf{I}_{n+1}(M; \mathbb{Z}_2)$, each of which is mapped by ∂ homeomorphically onto B_T.

Choose $\varepsilon \leq \mathrm{vol}(M)/3$ given by Lemma 2.2.2 and for every $T \in \mathcal{Z}_n(M; \mathbb{Z}_2)$ consider the open set

$$B_T = \{R \in \mathcal{Z}_n(M; \mathbb{Z}_2) : \mathcal{F}(T, R) < \varepsilon\}.$$

The family $\{B_T\}_{T \in \mathcal{Z}_n(M;\mathbb{Z}_2)}$ forms an open cover. With $\partial^{-1}T = \{U_1, U_2\}$ set

$$C_i = \{V \in \mathbf{I}_{n+1}(M; \mathbb{Z}_2) : \mathcal{F}(U_i, V) < \varepsilon\}, \quad i = 1, 2.$$

Note that $\mathcal{F}(U_1, U_2) = \mathbf{M}(U_1 - U_2) = \mathrm{vol}(M)$ and so C_1 and C_2 are disjoint. The reader can check that $C_1 \cup C_2 \subset \partial^{-1}(B_T)$. To check the reverse inclusion suppose that $R \in B_T$. From Lemma 2.2.2 we have the existence of $W \in \mathbf{I}_{n+1}(M; \mathbb{Z}_2)$ so that $\partial W = R - T$ and $\mathbf{M}(W) < \varepsilon$. Thus $V_i = W + U_i \in C_i$, $\partial V_i = R, i = 1, 2$, and so $C_1 \cup C_2 = \partial^{-1}(B_T)$. It also follows that each C_i is mapped homeomorphically to B_T for $i = 1, 2$, which proves the claim.

With S^k being a sphere of dimension k, consider a continuous map

$$\Psi : (S^k, *) \to (\mathcal{Z}_n(M; \mathbb{Z}_2), 0), \quad k \geq 2.$$

From the lifting criterion [12, Proposition 1.33] we have that the map Ψ admits a lift

$$\tilde{\Psi} : (S^k, *) \to (\mathbf{I}_{n+1}(M; \mathbb{Z}_2), 0)$$

because S^k is simply connected. From the fact that $\mathbf{I}_{n+1}(M; \mathbb{Z}_2)$ is contractible we obtain that Φ can be homotoped to the zero map. This proves

$$\pi_k(\mathcal{Z}_n(M; \mathbb{Z}_2), 0) = 0$$

for every $k \geq 2$. We now check that

$$\pi_1(\mathcal{Z}_n(M; \mathbb{Z}_2), 0) = \mathbb{Z}_2.$$

Given a loop γ in $\mathcal{Z}_{n+1}(M; \mathbb{Z}_2)$ with $\gamma(0) = \gamma(1) = 0$, the unique lifting property [12, Proposition 1.34] says there is a unique lift $\tilde{\gamma}$ to $\mathbf{I}_{n+1}(M; \mathbb{Z}_2)$ with $\tilde{\gamma}(0) = 0$. Thus, from Claim 2 one sees that the map

$$P : \pi_1(\mathcal{Z}_n(M; \mathbb{Z}_2), 0) \to \{0, M\}$$

which sends the homotopy class of γ to $\tilde{\gamma}(1)$ is well defined. The map is surjective because $\mathbf{I}_{n+1}(M; \mathbb{Z}_2)$ is path-connected and the reader can check that the map is injective.

Finally, we check that $\hat{\Phi}$ induces isomorphisms in every homotopy group. The curve

$$t \mapsto [\cos(\pi t) : \sin(\pi t) : 0 : \cdots], \quad 0 \leq t \leq 1,$$

generates $\pi_1(\mathbb{RP}^\infty, 1)$ and since the loop

$$\gamma(t) = \hat{\Phi}([\cos(\pi t) : \sin(\pi t) : 0 : \cdots]) = \partial\{f \leq -\cot(\pi t)\}, \quad 0 \leq t \leq 1$$

is homotopically non-trivial (because $P(\gamma) = M$), we deduce that the map

$$\hat{\Phi}_* : \pi_1(\mathbb{RP}^\infty, 1) \to \pi_1(\mathcal{Z}_n(M; \mathbb{Z}_2), 0)$$

is an isomorphism. The higher homotopy groups of both spaces are trivial, thus $\hat{\Phi}$ is a weak homotopy equivalence. $\qquad\square$

Theorem 2.2.3 and Hurewicz Theorem imply that

$$H^1(\mathcal{Z}_n(M; \mathbb{Z}_2); \mathbb{Z}_2) = \mathbb{Z}_2 = \{0, \bar{\lambda}\}.$$

We call $\bar{\lambda}$ the *fundamental cohomology class*. It has geometric meaning, namely, if $\sigma : S^1 \to \mathcal{Z}_n(M; \mathbb{Z}_2)$ is a loop then $\bar{\lambda}([\sigma]) = 1$ if and only if σ is homotopically non-trivial ($[\sigma]$ denotes the homology class induced by σ).

Let X denote a finite dimensional cubical subcomplex of some m-dimensional cube I^m. Every such cubical complex is homeomorphic to a finite simplicial complex and vice-versa (see Chapter 4 of [4]).

Definition 2.2.4 Let $k \in \mathbb{N}$. A continuous map $\Phi : X \to \mathcal{Z}_n(M; \mathbf{F}; \mathbb{Z}_2)$ is called a *k-sweepout* if $\lambda = \Phi^*(\bar{\lambda}) \in H^1(X, \mathbb{Z}_2)$ satisfies

$$\lambda^k = \lambda \smile \cdots \smile \lambda \neq 0 \in H^k(X, \mathbb{Z}_2),$$

where $\smile$ denotes the cup product.

The set of all k-sweepouts Φ is denoted by $\mathcal{P}_k$.

Remark In the definition above, the parameter space $X = \mathrm{dmn}(\Phi)$ of $\Phi \in \mathcal{P}_k$ is allowed to depend on Φ. Furthermore, every **F**-continuous map Φ' that is homotopic to Φ in the flat topology is also a k-sweepout.

We now argue that for all $k \in \mathbb{N}$ the set $\mathcal{P}_k$ is nonempty. The map $\Phi_k : \mathbb{RP}^k \to \mathcal{Z}_n(M; \mathbb{Z}_2)$ given by

$$\Phi_k([a_0 : a_1 : \cdots : a_k]) \mapsto \hat{\Phi}([a_0 : a_1 : \cdots : a_k : 0 : \cdots])$$

is such that $\lambda = \Phi_k^*(\bar{\lambda}) \neq 0$ in $H^1(\mathbb{RP}^k; \mathbb{Z}_2)$ and so $\lambda^k \neq 0$ in $H^k(\mathbb{RP}^k; \mathbb{Z}_2)$. Some work would be required to show that Φ_k is continuous in the **F**-metric and so instead we use Proposition 3.1 of [21] to find Ψ_k continuous in the **F**-topology and homotopic to Φ_k in the flat topology. Hence $\Psi_k \in \mathcal{P}_k$.

2.2.3 Min–Max Theorems

Let

$$\Phi : X \to \mathcal{Z}_n(M; \mathbf{F}; \mathbb{Z}_2)$$

be a continuous map. The *homotopy class* of Φ is the class Π of all continuous maps $\Phi' : X \to \mathcal{Z}_n(M; \mathbf{F}; \mathbb{Z}_2)$ such that Φ and Φ' are homotopic to each other in the flat topology.

If Φ is a k-sweepout then the corresponding homotopy class Π is non-trivial. Notice that our definition of homotopy class is slightly unusual, as we allow homotopies that are continuous in a weaker topology.

Definition 2.2.5 The *width* of Π is defined by:

$$\mathbf{L}(\Pi) = \inf_{\Phi \in \Pi} \sup\{\mathbf{M}(\Phi(x)) : x \in X\}.$$

It is implicitly assumed that every homotopy class Π being considered has $\mathbf{L}(\Pi) < \infty$.

Lemma 2.2.6 *If Π is a non-trivial homotopy class then $\mathbf{L}(\Pi) > 0$.*

Proof Consider $\varepsilon > 0$ given by Lemma 2.2.2. If $\mathbf{L}(\Pi) = 0$, we can find a map $\Phi \in \Pi$ so that Φ is a k-sweepout and $\mathbf{M}(\Phi(x)) < b_M^{-1} \varepsilon^{\frac{n}{n+1}}$ for all $x \in X = \mathrm{dmn}(\Phi)$, where b_M is the constant given by Federer–Fleming Isoperimetric Inequality. As a result we deduce from (2.2.1) that $\mathcal{F}(\Phi(x)) < \varepsilon$ for all $x \in X$. From Lemma 2.2.2 we have the existence of a unique $\Omega(x) \in \mathbf{I}_{n+1}(M; \mathbb{Z}_2)$ so that $\Phi(x) = \partial\Omega(x)$ and $\mathbf{M}(\Omega(x)) < \varepsilon$ for all $x \in X$, which means that Φ admits a lift $\tilde{\Phi}$ to $\mathbf{I}_{n+1}(M; \mathbb{Z}_2)$

that is continuous in flat topology. But $\mathbf{I}_{n+1}(M; \mathbb{Z}_2)$ is contractible and so the map Φ is homotopic to a constant map, which contradicts Π being non-trivial. $\square$

The following version of the Min–max Theorem, that follows from combining the existence theory of Almgren [2] and Pitts [24] with the regularity theory of Schoen and Simon [25], can be found in Section 3 of [20].

Min–Max Theorem 2.2.7 *Suppose* $\mathbf{L}(\Pi) > 0$. *There exists an integer n-varifold V with* $||V||(M) = \mathbf{L}(\Pi)$ *and support a closed minimal hypersurface that is smooth and embedded outside a set of Hausdorff dimension less than or equal to* $n - 7$.

This theorem has the following celebrated consequence (after combining with Lemma 2.2.6) which corresponds to Theorem 2.1.1.

Corollary 2.2.8 *Every closed Riemannian manifold* (M^{n+1}, g) *has a closed minimal hypersurface that is smooth and embedded outside a set of Hausdorff dimension less than or equal to* $n - 7$.

Consider a smooth embedded minimal cycle V so that $V = m_1|\Sigma_1| + \cdots + m_N|\Sigma_N|$ for a set $\{\Sigma_1, \ldots, \Sigma_N\}$ of closed, smooth, embedded, minimal hypersurfaces in M and a set $\{m_1, \ldots, m_N\} \subset \mathbb{N}$. The *Morse index* of V is the number

$$\text{index}(V) = \sum_{i=1}^{N} \text{index}(\Sigma_i).$$

If $m_1 = \cdots = m_N = 1$, we say V has *multiplicity one*.

From the definition of width, one sees that we maximize over a cubical complex X of dimension k and then minimize over an infinite dimensional space. Thus it is natural to expect that the Morse index of the smooth embedded minimal cycle should be bounded from above by k. This question was initially left unanswered in the original work of Almgren and Pitts. In [20] the authors showed that

Theorem 2.2.9 *Assume that* $3 \leq (n + 1) \leq 7$. *There exists a smooth embedded minimal cycle V so that*

$$||V||(M) = \mathbf{L}(\Pi) \quad and \quad \text{index}(V) \leq k.$$

We expect that a similar result should hold for dimensions higher than seven.

Lower bounds on the Morse index of smooth embedded minimal hypersurfaces is a subtler issue for the following reason: We can simply add some artificial parameters to the parameter space X so that we increase its dimension but the homotopy class of Π does not change. Thus, Morse index lower bounds have to be given in terms of the some topological property of Π rather than the dimension of the cubical complex X.

It turns out that obtaining optimal lower bounds for the Morse index is related with Multiplicity Once Conjecture made by authors in [21] (see also [20]) which states that

Multiplicity One Conjecture 2.2.10 *For generic metrics on M^{n+1}, $3 \leq (n+1) \leq$ 7, any component of a closed, minimal hypersurface obtained by min–max methods is two-sided and has multiplicity one.*

After this lectures were completed, Zhou [35, Theorem A] made a serious contribution to the min–max theory and, using a novel regularization of the area functional (developed by him and Zhu in [36]) proved the Multiplicity One Conjecture. Previously, in another tour-de-force, Chodosh and Mantoulidis [5] had proved this conjecture in the 3-dimensional case using the Allen–Cahn functional.

Multiplicity One Theorem 2.2.11 (Zhou) *Let (M^{n+1}, g) be a closed Riemannian manifold, $3 \leq (n+1) \leq 7$, with a bumpy metric.*

If Π is a non-trivial homotopy class there is an embedded, two-sided, multiplicity one, minimal hypersurface Σ with

$$\mathbf{L}(\Pi) = \text{vol}(\Sigma).$$

The result still holds if g is assumed simply to have positive Ricci curvature.

Remark Theorem A in [35] is stated assuming that (i) the homotopy class Π realizes the volume spectrum $\omega_k(M)$ (to be defined in Sect. 2.3.1) and that (ii) the maps in Π are defined on a cubical complex of dimension k. An inspection of the proof shows that (i) is not necessary and that (ii) can be dropped if one is not concerned about having sharp upper bounds on the Morse index of Σ that are also proven in [35, Theorem A].

The extension of the result to metrics of positive Ricci curvature is stated in [35, Remark 0.1] and the idea is to consider a sequence of bumpy metrics $\{g_i\}_{i \in \mathbb{N}}$ converging to g with $\text{Ric}(g) > 0$, apply Theorem A in [35] to obtain a sequence of embedded, two-sided, multiplicity one, minimal hypersurfaces Σ_i (with respect to metric g_i) and then use Sharp Compactness Theorem [26] to deduce the result for the metric g.

In [21], the authors showed optimal Morse index lower bounds assuming the Multiplicity One Conjecture. After Zhou's work we were able to remove that requirement (see [21, Addendum]) and showed

Theorem 2.2.12 *Let (M^{n+1}, g) be a closed Riemannian manifold, $3 \leq (n+1) \leq 7$ with a bumpy metric.*

Let Π be the homotopy class of a k-sweepout Φ defined on a k-dimensional cubical complex. There is an embedded, two-sided, multiplicity one, minimal hypersurface Σ with

$$\mathbf{L}(\Pi) = \text{vol}(\Sigma) \quad and \quad \text{index}(\Sigma) = k.$$

Remark In the Addendum of [21] the result is stated assuming that Π realizes the volume spectrum but that condition is not necessary.

2.3 Volume Spectrum and Weyl Law

Gromov [9] introduced the notion of volume spectrum, which will become extremely useful when paired with the min–max theory for minimal hypersurfaces.

2.3.1 Volume Spectrum

Recall the definition of k-sweepouts $\mathcal{P}_k$ given in Definition 2.2.4.

Definition 2.3.1 The *k-width* of (M, g) is defined to be

$$\omega_k(M, g) := \inf_{\Phi \in \mathcal{P}_k} \sup_{x \in \text{dmn}(\Phi)} \mathbf{M}(\Phi(x)).$$

The non-increasing sequence $\{\omega_k(M, g)\}_{k \in \mathbb{N}}$ is called the *volume spectrum* of (M, g).

When there is no risk of ambiguity, we denote the k-width simply by $\omega_k(M)$.

Remark Because of Proposition 3.1 in [21], the above definition of k-width coincides with the definition of k-width of [18] (Section 4.3) (or in [13, 16, 22]) where continuity in the **F**-metric in the definition of a k-sweepout is replaced by continuity in the flat topology together with a no concentration of mass property.

The following analogy with the Laplacian spectrum is instructive. The Rayleigh quotient is defined as

$$E : W^{1,2}(M) \setminus \{0\} \to [0, \infty), \quad E(f) = \frac{\int_M |\nabla f|^2 dV_g}{\int_M f^2 dV_g}$$

and the kth-eigenvalue $\lambda_k(M)$ of (M, g) is defined via the following min–max characterization:

$$\lambda_k(M) = \inf_{(k+1)-\text{plane } P} \max_{f \in P - \{0\}} E(f).$$

The Rayleigh quotient is scale invariant, meaning that $E(cf) = E(f)$ for all $c \neq 0$ and thus, considering the projectivization $\mathbb{P}W^{1,2}(M)$, where an element $[f] \in \mathbb{P}W^{1,2}(M)$, $f \neq 0$, represents the line $\{cf : c \in \mathbb{R}\} \subset W^{1,2}(M)$, we see that the

Rayleigh quotient descends to a map

$$E : \mathbb{P}W^{1,2}(M) \to [0, \infty), \quad E([f]) = \frac{\int_M |\nabla f|^2 dV_g}{\int_M f^2 dV_g}.$$

Note that $\mathbb{P}W^{1,2}(M)$ is homeomorphic to $\mathbb{R}\mathbb{P}^\infty$. In the same vein, a $(k + 1)$-plane in $W^{1,2}(M)$ projects to a k-dimensional projective subspace of $\mathbb{P}W^{1,2}(M)$ that we denote simply by $\mathbb{P}^k$ and thus $\lambda_k(M)$ is given by

$$\lambda_k(M) = \inf_{\mathbb{P}^k \subset \mathbb{P}W^{1,2}(M)} \max_{[f] \in \mathbb{P}^k} E([f]).$$

This identity has a striking similarity with Definition 2.3.1 and so, in that sense, $\{\omega_k(M)\}_{k \in \mathbb{N}}$ can be regarded as a non-linear spectrum.

It is worthwhile to point out that, unlike the spectrum for the Laplacian, the volume spectrum has not been computed on any specific example. Considering the unit 3-sphere S^3 with the standard metric, it is fairly straightforward to show that

$$\omega_1(S^3) = \omega_2(S^3) = \omega_3(S^3) = \omega_4(S^3) = \max_{\theta \in \mathbb{R}\mathbb{P}^4} \mathbf{M}(\Phi_4(\theta)) = 4\pi$$

but to show that $\omega_5(S^3) = \omega_6(S^3) = \omega_7(S^3) = 2\pi^2$, Nurser [23] had to use some of the work done by the authors in the solution to the Willmore conjecture [17], which should illustrate the subtleties of the problem. All other widths for S^3 are unknown.

The next result was essentially proven by Gromov [9, Section 4.2.B] and Guth [11]. A proof can also be found in Theorem 5.1 and Theorem 8.1 in [18].

Theorem 2.3.2 *There is a constant $C = C(M, g) > 0$ so that for all $k \in \mathbb{N}$*

$$C^{-1} k^{\frac{1}{n+1}} \le \omega_k(M) \le C k^{\frac{1}{n+1}}.$$

We postpone the proof of the lower bound to Corollary 2.3.7. Regarding the upper bound, we choose to present the different proof given in Theorem 3 of [3] which relies on a connection with the nodal sets of eigenfunctions that was made by authors in [18, Section 9].

Proof of Upper Bound Let $\bar{g}$ be an analytic metric on M and so we have $g \le c_1 \bar{g}$ for some constant c_1. With $\phi_0, \dots, \phi_p$ denoting the first $(p + 1)$-eigenfunctions for the Laplace operator of $(M, \bar{g})$, where ϕ_0 is the constant function, we can consider the map

$$\Phi_k : \mathbb{R}\mathbb{P}^p \to \mathcal{Z}_n(M; \mathbb{Z}_2),$$

$$\Phi_k([a_0, \dots, a_k]) = \partial\{x \in M : a_0\phi_0(x) + \dots + a_k\phi_k(x) < 0\}.$$

The map is well defined because we are considering mod 2 cycles and it was shown in [3] that the map is continuous in the flat topology and has no concentration of mass. This last part is relevant because we can then invoke Proposition 3.1 of [21] to find a map $\Psi_k \in \mathcal{P}_k$ so that

$$\sup_{y \in \mathbb{RP}^k} \mathbf{M}(\Psi_k(y)) \leq 2 \sup_{y \in \mathbb{RP}^k} \mathbf{M}(\Phi_k(y)).$$

Building on the volume estimates of nodal sets for analytic metrics of Donnely and Fefferman [6], it was shown in [14] that for some constant c_2,

$$\mathrm{vol}_{\tilde{g}}(\{x \in M : a_0\phi_0(x) + \ldots + a_k\phi_k(x) = 0\}) \leq c_2 k^{\frac{1}{n+1}}$$

and thus

$$\mathrm{vol}_g(\{x \in M : a_0\phi_0(x) + \ldots + a_k\phi_k(x) = 0\}) \leq c_1^{\frac{n}{2}} c_2 k^{\frac{1}{n+1}}$$

for all $k \in \mathbb{N}$ and $[a_0, \ldots, a_k] \in \mathbb{RP}^k$ so that $a_0\phi_0 + \ldots a_k\phi_k \neq 0$. Therefore

$$\sup_{y \in \mathbb{RP}^k} \mathbf{M}(\Psi_k(y)) \leq 2c_1^{\frac{n}{2}} c_2 k^{\frac{1}{n+1}} \quad \text{for all } k \in \mathbb{N}. \qquad \square$$

Consider the C^0-topology on the space of all metrics. The next proposition says that the map $g \mapsto k^{-\frac{1}{(n+1)}} \omega_k(M, g)$ is Lipschitz on sets of uniformly equivalent metrics, with a Lipschitz constant that does not depend on k.

Proposition 2.3.3 *Let $\tilde{g}$ be a Riemannian metric on M, and let c be a positive constant. Then there exists $K = K(\tilde{g}, c) > 0$ such that*

$$|k^{-\frac{1}{(n+1)}} \omega_k(M, g) - k^{-\frac{1}{(n+1)}} \omega_k(M, g')| \leq K \cdot |g - g'|_{\tilde{g}}$$

for any Riemannian metrics $c^{-1}\tilde{g} \leq g, g' \leq c\tilde{g}$ and any $k \in \mathbb{N}$.

Proof Given g, g' as above, we have

$$\sup_{v \neq 0} \frac{g'(v, v)}{g(v, v)} \leq 1 + \sup_{v \neq 0} \frac{|g(v, v) - g'(v, v)|}{g(v, v)} \leq 1 + c|g - g'|_{\tilde{g}}.$$

Assume $g \neq g'$ and choose a k-sweepout $\Phi : X \to \mathcal{Z}_n(M; ; \mathbf{F}; \mathbb{Z}_2)$ with

$$\sup\{\mathbf{M}_g(\Phi(x)) : x \in X\} \leq \omega_k(M, g) + |g - g'|_{\tilde{g}},$$

where $\mathbf{M}_g$ is the mass with respect to g. Then, considering the constant $C = C(M, \tilde{g})$ given by Theorem 2.3.2, we have

$$\omega_k(M, g') - \omega_k(M, g) \leq \sup\{\mathbf{M}_{g'}(\Phi(x)) : x \in X\} - \omega_k(M, g)$$

$$\leq \left(\sup_{v \neq 0} \frac{g'(v, v)}{g(v, v)}\right)^{\frac{n}{2}} \sup\{\mathbf{M}_g(\Phi(x)) : x \in X\} - \omega_k(M, g)$$

$$\leq \left(\sup_{v \neq 0} \frac{g'(v, v)}{g(v, v)}\right)^{\frac{n}{2}} (\omega_k(M, g) + |g - g'|_{\tilde{g}}) - \omega_k(M, g)$$

$$\leq ((1 + c|g - g'|_{\tilde{g}})^{\frac{n}{2}} - 1)\omega_k(M, g) + c^{\frac{n}{2}}|g - g'|_{\tilde{g}}$$

$$\leq ((1 + c|g - g'|_{\tilde{g}})^{\frac{n}{2}} - 1)c^{\frac{n}{2}}\omega_k(M, \tilde{g}) + c^{\frac{n}{2}}|g - g'|_{\tilde{g}}$$

$$\leq ((1 + c|g - g'|_{\tilde{g}})^{\frac{n}{2}} - 1)c^{\frac{n}{2}} Ck^{\frac{1}{(n+1)}} + c^{\frac{n}{2}}|g - g'|_{\tilde{g}},$$

from which the result follows.   □

2.3.2 Weyl Law for Volume Spectrum

A celebrated result concerning the spectrum of a manifold is the so called Weyl Law, which states that

$$\lim_{k \to \infty} \lambda_k(M)k^{-\frac{2}{n+1}} = a(n)\mathrm{vol}(M)^{-\frac{2}{n+1}},$$

where $a(n) = 4\pi^2\mathrm{vol}(B)^{-\frac{2}{n+1}}$ and B is the unit ball in $\mathbb{R}^{n+1}$. This was proven by Weyl [29] in 1911 for domains that are regions of space. The proof for closed manifolds came later in 1949, by Minakshisundaram and Pleijel, and uses the asymptotic expansion for the trace of the heat kernel.

Gromov conjectured ([10, 8.4]) that the volume spectrum $\{\omega_p(M)\}_{p \in \mathbb{N}}$ satisfies a Weyl's asymptotic law. Jointly with Liokumovich, the authors confirmed this conjecture and showed in [16] the following result.

Weyl Law for the Volume Spectrum 2.3.4 *There exists a constant $a(n) > 0$ such that, for every compact Riemannian manifold (M^{n+1}, g) with (possibly empty) boundary, we have*

$$\lim_{k \to \infty} \omega_k(M)k^{-\frac{1}{n+1}} = a(n)\mathrm{vol}(M)^{\frac{n}{n+1}}.$$

Before sketching its proof it is worthwhile to make some comments. Unlike the spectrum of the Laplacian that is known in several cases (like round spheres or cubes) the volume spectrum, due in part to being a non-linear spectrum, has not been computed on any specific example. Nonetheless, we were able to prove a universal asymptotic law without knowing the value of the universal constant $a(n)$, which is in stark contrast with both Weyl and Minakshisundaram–Pleijel proofs. Moreover, Minakshisundaram–Pleijel proof for closed manifolds uses techniques that do not seem to have an analogue for the volume spectrum and so a new approach had to be developed.

In order to prove the Weyl Law we need to introduce relative cycles and mention their basic features.

Let (Ω, g) be Riemannian compact $(n+1)$-manifold with Lipschitz boundary $\partial\Omega$ and $H_{n+1}(\Omega, \partial\Omega, \mathbb{Z}_2) = \mathbb{Z}_2$. We denoted them by *connected Lipschitz domains*.

Consider the space

$$\mathbf{I}_n(\Omega, \partial\Omega; \mathbb{Z}_2) = \{T \in \mathbf{I}_n(\Omega; \mathbb{Z}_2) : \text{support}(\partial T) \subset \partial\Omega\}.$$

We say that $T, S \in \mathbf{I}_n(\Omega, \partial\Omega; \mathbb{Z}_2)$ are equivalent if $T - S \in \mathbf{I}_n(\partial\Omega; \mathbb{Z}_2)$ and the connected component containing zero of the space of such equivalence classes, called *mod 2 relative n-cycles*, is denoted by $\mathcal{Z}_n(\Omega, \partial\Omega; \mathbb{Z}_2)$. We abuse notation and use $T \in \mathbf{I}_n(\Omega, \partial\Omega; \mathbb{Z}_2)$ to denote its equivalence class as a mod 2 relative n-cycle. In [16, Section 2.2] a further subscript appears in the notation of mod 2 relative n-cycles.

The mass, flat metric, and **F**-metric on $\mathcal{Z}_n(\Omega, \partial\Omega; \mathbb{Z}_2)$ are defined as

$$\mathbf{M}(T) = \inf\{\mathbf{M}(T + R) : R \in \mathbf{I}_n(\partial\Omega; \mathbb{Z}_2)\},$$

$$\mathcal{F}(S, T) = \inf\{\mathcal{F}(S + R, T) : R \in \mathbf{I}_n(\partial\Omega; \mathbb{Z}_2)\}$$

and

$$\mathbf{F}(S, T) = \inf\{\mathbf{F}(S + R, T), \mathbf{F}(T + R, S) : R \in \mathbf{I}_n(\partial\Omega; \mathbb{Z}_2)\}$$

for all S, T in $\mathcal{Z}_n(\Omega, \partial\Omega; \mathbb{Z}_2)$. These metrics induced the flat and **F**-topology, respectively.

The theory for $\mathcal{Z}_n(\Omega, \partial\Omega; \mathbb{Z}_2)$ mimics the theory for $\mathcal{Z}_n(M; \mathbb{Z}_2)$ (see [16, Section 2] for details). Namely,

$$H^1(\mathcal{Z}_n(\Omega, \partial\Omega; \mathbb{Z}_2); \mathbb{Z}_2) = \mathbb{Z}_2 = \{0, \bar{\lambda}\},$$

and the set of k-sweepouts $\mathcal{P}_k$ is defined as in Definition 2.2.4. One has $\mathcal{P}_k \neq \emptyset$ for all $k \in \mathbb{N}$ and the k-width $\omega_k(\Omega)$ is defined exactly like in Definition 2.3.1. In [16], k-sweepouts and k-width are defined in terms of maps that are continuous in the flat topology and have no concentration of mass but using the approximation results of [16, Section 2.9] one can show that the value for $\omega_k(\Omega)$ remains the same if one requires the maps to be continuous in the **F**-topology instead. Finally, similarly to

Theorem 2.3.2, there is a constant $C = C(\Omega, g)$ so that

$$\omega_k(\Omega) \leq C k^{\frac{1}{n+1}} \quad \text{for all } k \in \mathbb{N}. \tag{2.3.2}$$

We first prove Theorem 2.3.4 for connected Lipschitz domains using a superadditivity property for the k-widths. The key ingredients are the min–max definition of k-width and the vanishing property of the cup product. Weyl's proof is also based on similar properties for the Laplacian eigenvalues.

Lusternick–Schnirelman Superadditivity 2.3.5 *Consider connected Lipschitz domains Ω_0, $\{\Omega_i^*\}_{i=1}^N$ such that*

- $\Omega_i^* \subset \Omega_0$ *for all $i = 1, \ldots, N$ and the interiors of $\{\Omega_i^*\}_{i=1}^N$ are pairwise disjoint.*

Then, given positive integers so that $k_i + \ldots + k_N \leq k$, we have

$$\omega_k(\Omega_0) \geq \sum_{i=1}^N \omega_{k_i}(\Omega_i^*).$$

Sketch of Proof Set $\bar{k} = \sum_{i=1}^N k_i$. Given Φ a k-sweepout of Ω_0 (with $X = \mathrm{dmn}(\Phi)$) and $\lambda = \Phi^* \bar{\lambda} \in H^1(X, \mathbb{Z}_2)$, we assume for simplicity that the set

$$U_i = \{x \in X : \mathbf{M}(\Phi(x) \llcorner \Omega_i^*) < \omega_{k_i}(\Omega_i^*)\}$$

is open and the map

$$\Phi_i : X \to \mathcal{Z}_n(\Omega_i^*, \partial\Omega_i^*; \mathbf{F}; \mathbb{Z}_2) \quad x \mapsto \Phi(x) \llcorner \Omega_i^*$$

is well defined for all $1 \leq i \leq N$. The general argument can be found in Theorem 3.1 of [16].

Fix $1 \leq i \leq N$. With $\iota : U_i \to X$ the inclusion map, we have from the definition of U_i that $\Phi_i \circ \iota$ is not a k_i-sweepout of Ω_i, which means that $\iota^* \lambda^{k_i} = (\Phi_i \circ \iota)^* \bar{\lambda}^{k_i} = 0$ in $H^{k_i}(U_i; \mathbb{Z}_2)$. Therefore λ^{k_i} vanishes on U_i for all $1 \leq i \leq N$. The vanishing property for the cup product [12, page 209] implies that

$$\lambda^{\bar{k}} = \lambda^{k_1} \smile \ldots \smile \lambda^{k_N}$$

vanishes on $\cup_{i=1}^N U_i$. But $\lambda^{\bar{k}} \neq 0$ on X because $\bar{k} \leq k$ and so $X \neq \cup_{i=1}^N U_i$.

Choose $x \in X \setminus \cup_{i=1}^N U_i$. Combining the definition of U_i with the fact that the interiors of $\{\Omega_i^*\}_{i=1}^N$ are pairwise disjoint we have that

$$\Phi(x) \geq \sum_{i=1}^N \mathbf{M}(\Phi(x) \llcorner \Omega_i^*) \geq \sum_{i=1}^N \omega_{k_i}(\Omega_i^*). \qquad \square$$

We use $\lfloor x \rfloor$ to denote the integer part.

Corollary 2.3.6 *Under the same conditions of the Lusternick–Schnirelman Super-additivity, assume also that*

- Ω_0 *has unit volume and* $\Omega_0, \{\Omega_i^*\}_{i=1}^N \subset \mathbb{R}^{n+1}$.

For all $i = 1, \ldots, N$, *denote by* Ω_i *a scaling of* Ω_i^* *with unit volume.*

Then, with $V = \min\{\mathrm{vol}(\Omega_i^*)\}_{i=1}^N$ *and* $k_i = \lfloor k\mathrm{vol}(\Omega_i^*) \rfloor$, $i = 1, \ldots, N$, *we have for all* $k \in \mathbb{N}$

$$k^{-\frac{1}{n+1}} \omega_k(\Omega_0) \geq \sum_{i=1}^N \mathrm{vol}(\Omega_i^*) k_i^{-\frac{1}{n+1}} \omega_{k_i}(\Omega_i) + O\left(\frac{1}{kV}\right).$$

Proof The p-width scales like n-dimensional area and so we have $\omega_p(\Omega_i^*) = \mathrm{vol}(\Omega_i^*)^{\frac{n}{n+1}} \omega_p(\Omega_i)$ for all $i = 1, \ldots, N$ and $p \in \mathbb{N}$.

We have $\sum_{i=1}^N k_i \leq k\mathrm{vol}(\Omega_0) = k$ and so, using Lusternick–Schnirelman Superadditivity and (2.3.2), we deduce

$$k^{-\frac{1}{n+1}} \omega_k(\Omega_0) \geq k^{-\frac{1}{n+1}} \sum_{i=1}^N \omega_{k_i}(\Omega_i^*)$$

$$= k^{-\frac{1}{n+1}} \sum_{i=1}^N \mathrm{vol}(\Omega_i^*)^{\frac{n}{n+1}} \omega_{k_i}(\Omega_i)$$

$$= \sum_{i=1}^N \mathrm{vol}(\Omega_i^*) \left(\frac{k_i}{k\mathrm{vol}(\Omega_i^*)}\right)^{\frac{1}{n+1}} k_i^{-\frac{1}{n+1}} \omega_{k_i}(\Omega_i)$$

$$\geq \sum_{i=1}^N \mathrm{vol}(\Omega_i^*) \left(1 - \frac{1}{k\mathrm{vol}(\Omega_i^*)}\right)^{\frac{1}{n+1}} k_i^{-\frac{1}{n+1}} \omega_{k_i}(\Omega_i)$$

$$= \sum_{i=1}^N \mathrm{vol}(\Omega_i^*) k_i^{-\frac{1}{n+1}} \omega_{k_i}(\Omega_i) + O\left(\frac{1}{kV}\right). \qquad \square$$

From the Lusternick–Schnirelman Superadditivity 2.3.5 we can also deduce the lower bounds for the k-width stated in Theorem 2.3.2.

Corollary 2.3.7 *There is a constant* $C = C(M, g) > 0$ *so that for all* $k \in \mathbb{N}$

$$\omega_k(M) \geq C^{-1} k^{\frac{1}{n+1}}.$$

Proof Given $p \in M$, let $B_r(p)$ denote the geodesic ball in M of radius r and centered at p, and consider $\omega_1(B)$, where B is the unit ball in $\mathbb{R}^{n+1}$. Lemma 2.2.6 extends to the context of relative cycles to conclude that $\omega_1(B) > 0$.

There is $\bar{r}$ small so that for all $r \leq \bar{r}$ and $p \in M$ we have

$$\omega_1(B_r(p), g) \geq r^n \omega_1(B)/2.$$

Moreover, there exists some constant $\nu = \nu(M) > 0$ such that, for every $k \in \mathbb{N}$, one can find a collection of k disjoint geodesic balls $\{B_j\}_{j=1}^k$ of radius $r = \nu k^{-\frac{1}{n+1}}$. Hence, we deduce from Lusternick–Schnirelman Superadditivity 2.3.5

$$\omega_k(M, g) \geq \sum_{j=1}^k \omega_1(B_j, g) \geq kr^n \frac{\omega_1(B)}{2} = kk^{-\frac{n}{n+1}} \nu^n \frac{\omega_1(B)}{2} = k^{\frac{1}{n+1}} \nu^n \frac{\omega_1(B)}{2}. \quad \square$$

We are now ready to show

Weyl Law for Domains 2.3.8 *Let $\Omega \subset \mathbb{R}^{n+1}$ be a connected Lipschitz domain $\Omega \subset \mathbb{R}^{n+1}$. There is a universal constant $a(n)$ so that for every $\Omega \subset \mathbb{R}^{n+1}$ we have*

$$\lim_{k \to \infty} k^{-\frac{1}{n+1}} \omega_k(\Omega) = a(n) \mathrm{vol}(\Omega)^{\frac{1}{n+1}}.$$

In [16, Theorem 3.2] this result was also proven for higher codimension relative cycles.

Proof Without loss of generality we assume that $\mathrm{vol}(\Omega) = 1$. Let C denote the unit cube in $\mathbb{R}^{n+1}$ and set

$$a_-(n) = \liminf_{k \to \infty} k^{-\frac{1}{n+1}} \omega_k(C) \quad \text{and} \quad a^+(n) = \limsup_{p \to \infty} k^{-\frac{1}{n+1}} \omega_k(C).$$

Claim 1 $a_-(n) = a^+(n)$ *and so define* $a(n)$ *to be that common value.*
  Choose $\{k_l\}_{l \in \mathbb{N}}$, $\{p_j\}_{j \in \mathbb{N}}$ so that

$$\lim_{l \to \infty} k_l^{-\frac{1}{n+1}} \omega_{k_l}(C) = a^+(n) \quad \text{and} \quad \lim_{j \to \infty} p_j^{-\frac{1}{n+1}} \omega_{p_j}(C) = a_-(n).$$

With l fixed and for all j large enough so that $\delta_j := k_l/p_j < 1$, consider N_j to be the maximum number of cubes $\{C_i^*\}_{i=1}^{N_j}$ with pairwise disjoint interiors contained in C and all with the same volume δ_j. We must have $\delta_j N_j \to 1$ as $j \to \infty$. From Corollary 2.3.6 we obtain

$$p_j^{-\frac{1}{n+1}} \omega_{p_j}(C) \geq \sum_{i=1}^{N_j} \mathrm{vol}(C_i^*) k_l^{-\frac{1}{n+1}} \omega_{k_l}(C) + O(k_l^{-1})$$

$$= \delta_j N_j k_l^{-\frac{1}{n+1}} \omega_{k_l}(C) + O(k_l^{-1}).$$

Making $j \to \infty$ and then $l \to \infty$ we deduce the claim.

Claim 2 $\liminf_{k\to\infty} k^{-\frac{1}{n+1}} \omega_k(\Omega) \geq a(n)$.

Given any $\varepsilon > 0$, one can find a family of cubes $\{C_i^*\}_{i=1}^N$ with pairwise disjoint interiors contained in Ω, all with the same volume δ, and such that

$$\sum_{i=1}^N \text{vol}(C_i^*) \geq 1 - \varepsilon.$$

From Corollray 2.3.6 we obtain, with $k_\delta = \lfloor k\delta \rfloor$,

$$k^{-\frac{1}{n+1}} \omega_k(\Omega) \geq \sum_{i=1}^N \text{vol}(C_i^*) k_\delta^{-\frac{1}{n+1}} \omega_{k_\delta}(C) + O\left(\frac{1}{k\delta}\right)$$

and thus making $k \to \infty$ we have

$$\liminf_{k\to\infty} k^{-\frac{1}{n+1}} \omega_k(\Omega) \geq (1 - \varepsilon) \liminf_{k\to\infty} k^{-\frac{1}{n+1}} \omega_k(C) = (1 - \varepsilon)a(n).$$

The claim follows from the arbitrariness of ε.

Claim 3 $\limsup_{k\to\infty} k^{-\frac{1}{n+1}} \omega_k(\Omega) \leq a(n)$.

Define $\tilde{\omega}_k(U) := k^{-\frac{1}{n+1}} \omega_k(U)$ for every connected Lipschitz domain U. Choose $\{k_j\}_{j\in\mathbb{N}}$ so that

$$\beta := \lim_{j\to\infty} \tilde{\omega}_{k_j}(\Omega) = \limsup_{k\to\infty} \tilde{\omega}_k(\Omega).$$

Choose $\varepsilon > 0$. From Lemma 3.5 in [16] we can choose domains $\{\Omega_i^*\}_{i=1}^N$ contained in C so that (i) every Ω_i^* is a scaling of Ω, (ii) their interiors are pairwise disjoint, and (iii) $\sum_{i=1}^N \text{vol}(\Omega_i^*) \geq 1 - \varepsilon$.

Fix $j \in \mathbb{N}$ and pick $p_j \in \mathbb{N}$ so that $\lfloor p_j \text{vol}(\Omega_1^*) \rfloor = k_j$ (such choice is possible because $\text{vol}(\Omega_1^*) \leq 1$). With $V_N = \min\{\text{vol}(\Omega_i^*)\}_{i=1}^N$ and $k_{i,j} = \lfloor p_j vol(\Omega_i^*) \rfloor$, we have from Corollary 2.3.6 that

$$\tilde{\omega}_{p_j}(C) \geq \sum_{i=1}^N \text{vol}(\Omega_i^*) \tilde{\omega}_{k_{i,j}}(\Omega) + O\left(\frac{1}{p_j V_N}\right)$$

$$= \text{vol}(\Omega_1^*) \tilde{\omega}_{k_j}(\Omega) + \sum_{i=2}^N \text{vol}(\Omega_i^*) \tilde{\omega}_{k_{i,j}}(\Omega) + O\left(\frac{1}{p_j V_N}\right).$$

Making $j \to \infty$ and using Claim 2 we have

$$a(n) \geq \text{vol}(\Omega_1^*)\beta + a(n) \sum_{i=2}^N \text{vol}(\Omega_i^*) \geq \text{vol}(\Omega_1^*)\beta + a(n)(1 - \varepsilon - \text{vol}(\Omega_1^*)).$$

Making $\varepsilon \to 0$ we obtain

$$a(n) \geq \text{vol}(\Omega_1^*)\beta + a(n)(1 - \text{vol}(\Omega_1^*)),$$

and so $a(n) \geq \beta$, which finishes the claim. □

We now explain the key ideas to show the theorem below

Weyl Law for Compact Manifolds 2.3.9 *For every closed Riemannian manifold (M^{n+1}, g) with (possibly empty) boundary, we have*

$$\lim_{k\to\infty} \omega_k(M)k^{-\frac{1}{n+1}} = a(n)\text{vol}(M)^{\frac{n}{n+1}}.$$

Sketch of Proof Without loss of generality we assume that $\text{vol}(M) = 1$. We also assume that $\partial M = \emptyset$ for simplicity.

The idea to show

$$\lim_{k\to\infty} k^{-\frac{1}{n+1}}\omega_k(M) \geq a(n)$$

is the following: We find a sufficiently large number of small pairwise disjoint geodesic balls $\{B_i\}_{i=1}^N \subset M$ so that $\sum_{i=1}^N \text{vol}(B_i) \simeq 1$ and the metric g on B_i is close to being the Euclidean metric on a ball. Due to the Weyl Law for domains 2.3.8, we have $\omega_k(B_i)k^{-\frac{1}{n+1}} \simeq a(n)\text{vol}(B_i)^{\frac{n}{n+1}}$ for all k very large. Thus from the Lusternick–Schnirelman Superadditivity 2.3.5 we deduce, for all k sufficiently large,

$$k^{-\frac{1}{n+1}}\omega_k(M) \geq k^{-\frac{1}{n+1}}\sum_{i=1}^N \omega_{\lfloor k\text{vol}(B_i)\rfloor}(B_i) = \sum_{i=1}^N k^{-\frac{1}{n+1}}\omega_{\lfloor k\text{vol}(B_i)\rfloor}(B_i)$$

$$\simeq \sum_{i=1}^N a(n)\text{vol}(B_i) \simeq a(n).$$

The reader can see the details in Theorem 4.1 of [16].

To prove the other inequality, the first step consists in decomposing M into regions $\{C_i\}_{i=1}^N$ so that:

- Each C_i is $(1+\varepsilon)$-bi-Lipschitz diffeomorphic to a Lipschitz domain C_i in $\mathbb{R}^{n+1}$;
- The regions $\{C_i\}_{i=1}^N$ cover M;
- $\{C_i\}_{i=1}^N$ and $\{C_i\}_{i=1}^N$ have mutually disjoint interiors, respectively.

We then connect the disjoint regions $\{C_i\}_{i=1}^N \subset \mathbb{R}^{n+1}$ with tubes of very small volume so that we obtain a connected Lipschitz domain Ω. By making ε smaller, we can make the volume of Ω arbitrarily close to $\text{vol}(M)$.

In what follows we will be content with producing sweepouts that are only continuous with respect to the flat topology (instead of continuous with respect to the **F**-topology). The reader can see the general argument in [16, Theorem 4.2].

Consider Φ a k-sweepout of Ω with $X = \text{dmn}(\Phi)$, which then induces k-sweepouts on each C_i given by

$$\Phi_i : X \to \mathcal{Z}_n(C_i, \partial C_i; \mathbb{Z}_2), \quad \Phi_i(x) = \Phi(x) \llcorner C_i, \quad i = 1, \dots, N.$$

In [16, Section 4] we show that, after a possibly small perturbation, the map Φ_i is well defined and a k-sweepout with

$$\lambda := \Phi_i^* \bar{\lambda} = \Phi^* \bar{\lambda} \quad \text{for all } i = 1, \dots, N.$$

The general idea is to use the maps $\{\Phi_i\}_{i=1}^N$ to construct a k-sweepout of M as follows: For every $x \in X$ the elements $\Phi_i(x)$ have boundary in ∂C_i and we show in [16, Lemma 4.3] the existence of $Z_i(x) \in \mathbf{I}_{n+1}(C_i; \mathbb{Z}_2)$ so that the cycle $\partial Z_i(x)$ coincides with $\Phi_i(x)$ on the interior of C_i. Because the choice of $Z_i(x)$ is not unique ($C_i + Z_i(x)$ would have also been a valid choice) it is not always possible to construct a continuous map $x \mapsto \partial Z_i(x)$. Nonetheless, we will argue that a choice of $Z_1(x)$ for a given x will induce choices of $Z_2(x), \dots, Z_N(x)$ so that if $\tilde{Z}_i(x)$ denotes the image of $Z_i(x)$ in C_i under the respective bi-Lipschitz diffeomorphism, then $\partial \tilde{Z}_1(x) + \dots + \partial \tilde{Z}_N(x)$ is a cycle in M that does not depend on the choice of $Z_1(x)$ and we use that to conclude that the map $x \mapsto (\partial \tilde{Z}_1 + \dots + \partial \tilde{Z}_N)(x)$ is continuous. We now provide some of the details.

For each $i = 1, \dots, N$ set

$$SX_i = \{(x, Z) : x \in X, \Phi_i(x) - \partial Z \in \mathbf{I}_n(\partial C_i; \mathbb{Z}_2)\} \subset X \times \mathbf{I}_{n+1}(C_i; \mathbb{Z}_2).$$

From the Constancy Theorem we have that $\mathbf{I}_n(\partial C_i; \mathbb{Z}_2) = \{0, \partial C_i\}$ for all $i = 1, \dots, N$. Thus if $(x, Z), (x, Z') \in SX_i$ we have that either $Z = Z'$ or $Z = C_i - Z'$. There is a natural projection $\tau_i : SX_i \to X$ and in Lemma 4.3 of [16] we show that τ_i is a 2-cover of X for all $i = 1, \dots, N$.

Claim 1 SX_1 is isomorphic to SX_i for all $i = 1, \dots, N$.

The isomorphism classes of double covers of X are in a bijective correspondence with $\text{Hom}(\pi_1(X), \mathbb{Z}_2)$, which is homeomorphic to $H^1(X; \mathbb{Z}_2)$. It suffices to see that, for all $i = 1, \dots, N$, the element $\sigma_i \in H^1(X; \mathbb{Z}_2)$ that classifies SX_i is identical to λ. Indeed given $\gamma : S^1 \to X$ nontrivial in $\pi_1(X)$, consider a lift to SX_i given by $\theta \mapsto (\gamma(\exp(i\theta)), Z_\theta), 0 \le \theta \le 2\pi$. Then $\sigma_i(\gamma)$ is 1 if $Z_0 = C_i - Z_{2\pi}$ and 0 if $Z_0 = Z_{2\pi}$. Thus $\sigma_i(\gamma)$ is non-zero if and only if $\Phi_i \circ \gamma$ is a sweepout.

As a result we obtain that SX_1 is isomorphic to SX_i for all $i = 1, \dots, N$ and let $F_i : SX_1 \to SX_i$ be the corresponding isomorphism. Given an element $v = (x, T)$ in SX_i, we denote by $\Xi_i(v) \in \mathbf{I}_{n+1}(C_i; \mathbb{Z}_2)$ the image of T under the bi-Lipschitz diffeomorphism from C_i to C_i, $i = 1, \dots, N$. Using this notation we consider the continuous map in the flat topology

$$\hat{\Psi} : SX_1 \to \mathcal{Z}_n(M; \mathbb{Z}_2), \quad \hat{\Psi}(y) = \sum_{i=1}^{N} \partial\, \Xi_i(F_i(y)).$$

If $(x, Z) \in SX_1$, then $\Xi_i(F_i(x, C_1 + Z)) = C_i + \Xi_i(F_i(x, Z))$ for all $i = 1, \ldots, N$, and so

$$\hat{\Psi}(x, C_1 + Z) = \sum_{i=1}^{N} \partial(C_i + \Xi_i(F_i(x, Z))) = \sum_{i=1}^{N} \partial C_i + \hat{\Psi}(x, Z)$$

$$= \partial M + \hat{\Psi}(x, Z) = \hat{\Psi}(x, Z).$$

Thus $\hat{\Psi}(x, C_1 + Z) = \hat{\Psi}(x, Z)$ in $\mathcal{Z}_{n,}(M; \mathbb{Z}_2)$, which means that $\hat{\Psi}$ descends to a continuous map in the flat topology $\Psi : X \to \mathcal{Z}_n(M; \mathbb{Z}_2)$.

Claim 2 Ψ is a p-sweepout.

Choose $\gamma : S^1 \to X$ nontrivial in $\pi_1(X)$ and denote by γ_1 its lift to SX_1. Then $\gamma_i = F_i \circ \gamma_1$ gives a lift to SX_i for all $i = 1, \ldots, N$ and we consider the continuous map in the flat topology

$$B : [0, 2\pi] \to \mathbf{I}_{n+1}(M; \mathbb{Z}_2), \quad B(\theta) = \sum_{i=1}^{N} \Xi_i(\gamma_i(\theta)).$$

We have $(\Psi \circ \gamma)(\theta) = \partial B(\theta)$ for all $0 \le \theta \le 2\pi$.

Hence $\Psi^*\bar{\lambda} = \lambda$ because, recalling that $\sigma_i = \lambda$ for all $i = 1, \ldots, N$,

$$\lambda(\gamma) = 0 \implies \sigma_i(\gamma) = 0 \text{ for all } i = 1, \ldots, N$$
$$\implies \Xi_i(\gamma_i(2\pi)) = \Xi_i(\gamma_i(0)) \text{ for all } i = 1, \ldots, N$$
$$\implies B(2\pi) = B(0)$$

and

$$\lambda(\gamma) = 1 \implies \sigma_i(\gamma) = 1 \text{ for all } i = 1, \ldots, N$$
$$\implies \Xi_i(\gamma_i(2\pi)) = C_i + \Xi_i(\gamma_i(0)) \text{ for all } i = 1, \ldots, N$$
$$\implies B(2\pi) = M + B(0),$$

where in the last line we used the fact that $\{C_i\}_{i=1}^{N}$ are pairwise disjoint and cover M. This implies that Ψ is a p-sweepout because $\lambda^p \ne 0$.

Throughout the rest of the proof we ignore the ε-dependence in some of the constants for simplicity.

Claim 3 For all $x \in X$ we have

$$\mathbf{M}(\Psi(x)) \lesssim \mathbf{M}(\Phi(x)) + \sum_{i=1}^{N} \mathrm{vol}(\partial C_i).$$

Given $(x, Z) \in SX_i, i = 1, \ldots, N$, we have that

$$\mathbf{M}(\Xi_i(\partial Z)) \simeq \mathbf{M}(\partial Z) \leq \mathbf{M}(\Phi_i(x)) + \mathbf{M}(\partial C_i)$$

Hence, recalling the definition of the maps Φ_i, we have

$$\mathbf{M}(\Psi(x)) \leq \sum_{i=1}^{N} \mathbf{M}(\partial \, \Xi_i(F_i(y))) \lesssim \sum_{i=1}^{N} (\mathbf{M}(\Phi_i(x)) + \mathbf{M}(\partial C_i))$$

$$\leq \sum_{i=1}^{N} \mathbf{M}(\Phi(x) \llcorner C_i) + \sum_{i=1}^{N} \mathrm{vol}(\partial C_i) \leq \mathbf{M}(\Phi(x)) + \sum_{i=1}^{N} \mathrm{vol}(\partial C_i).$$

Combining Claims 2 with 3 we deduce that for all $k \in \mathbb{N}$

$$\omega_k(M) \lesssim \omega_k(\Omega) + \sum_{i=1}^{N} \mathrm{vol}(\partial C_i).$$

From Theorem 2.3.8 we have that

$$\limsup_{k \to \infty} k^{-\frac{1}{n+1}} \omega_k(M) \lesssim \lim_{k \to \infty} k^{-\frac{1}{n+1}} \omega_k(\Omega) \leq a(n) \mathrm{vol}(\Omega)^{\frac{n}{n+1}} \simeq a(n) \mathrm{vol}(M)^{\frac{n}{n+1}}.$$

$\qquad\qquad\qquad\qquad\qquad\qquad\qquad\qquad\qquad\qquad\qquad\qquad\qquad\qquad\qquad\qquad\qquad\square$

2.3.3 Min–Max Theory and the Volume Spectrum

Combining the Weyl Law for the Volume Spectrum 2.3.4 with Theorem 2.2.9 and Theorem 2.2.12 we obtain

Theorem 2.3.10 *Assume (M^{n+1}, g) is a closed Riemannian manifold, $3 \leq (n + 1) \leq 7$.*

For each $k \in \mathbb{N}$ there exist a smooth embedded cycle V so that

$$\mathrm{vol}(V) = \omega_k(M) \simeq a(n) \mathrm{vol}(M)^{\frac{n}{n+1}} k^{\frac{1}{n+1}} \quad and \quad \mathrm{index}(V) \leq k,$$

where $a(n)$ is a universal constant.

If the metric g is bumpy there is an embedded, two-sided, multiplicity one, minimal hypersurface Σ_k with

$$\text{vol}(\Sigma_k) = \omega_k(M) \simeq a(n)\text{vol}(M)^{\frac{n}{n+1}} k^{\frac{1}{n+1}} \quad and \quad \text{index}(\Sigma_k) = k.$$

Proof Choose a sequence $\{\Phi_i\}_{i\in\mathbb{N}} \subset \mathcal{P}_k$ such that

$$\lim_{i\to\infty} \sup\{\mathbf{M}(\Phi_i(x)) : x \in X_i = \text{dmn}(\Phi_i)\} = \omega_k(M).$$

Denote by $X_i^{(k)}$ the k-dimensional skeleton of X_i. Then $H^k(X_i, X_i^{(k)}; \mathbb{Z}_2) = 0$ and hence the long exact cohomology sequence gives that the natural pullback map from $H^k(X_i; \mathbb{Z}_2)$ into $H^k(X_i^{(k)}; \mathbb{Z}_2)$ is injective. This implies $(\Phi_i)_{|X_i^{(k)}} \in \mathcal{P}_k$. The definition of $\omega_k(M)$ then implies

$$\lim_{i\to\infty} \sup\{\mathbf{M}(\Phi_i(x)) : x \in X_i^{(k)}\} = \omega_k(M).$$

We denote by Π_i the homotopy class of $(\Phi_i)_{|X_i^{(k)}}$. Its width $\mathbf{L}(\Pi_i)$ satisfies

$$\omega_k(M) \leq \mathbf{L}(\Pi_i) \leq \sup\{M(\Phi_i(x)) : x \in X_i^{(k)}\}, \quad i \in \mathbb{N}$$

and in particular $\lim_{i\to\infty} \mathbf{L}(\Pi_i) = \omega_k(M)$.

Theorem 2.2.9 implies, for all $i \in \mathbb{N}$, the existence of smooth embedded cycles V_i so that

$$\mathbf{L}(\Pi_i) = \text{vol}(V_i) \quad and \quad \text{index}(V_i) \leq k.$$

The Compactness Theorem of Sharp (Theorem 2.3 of [26]) gives the existence of a smooth embedded cycles V with $\text{index}(V) \leq k$ such that, after passing to a subsequence, $\text{vol}(V_i) \to \text{vol}(V)$ as $i \to \infty$, which finishes the proof in the general case.

When g is bumpy we have from Sharp (Theorem 2.3 and Remark 2.4, [26]) that the set of connected, closed, smooth, embedded minimal hypersurfaces in (M, g) with both area and index uniformly bounded is finite and so there must exist some $j \in \mathbb{N}$ so that $\omega_k(M) = \mathbf{L}(\Pi_j)$. The result follows from Theorem 2.2.12.   □

This theorem, when combined with the Multiplicity One Theorem 2.2.11 has the following corollary and corresponds to Theorem B in [35].

Corollary 2.3.11 *Assume (M^{n+1}, g) is a closed Riemannian manifold, $3 \leq (n + 1) \leq 7$ with either a bumpy metric or a metric with positive Ricci curvature. Then there exists infinitely many smooth, connected, closed, embedded, minimal hypersurfaces.*

The argument used to prove Theorem 2.1.4 in [18] used a different reasoning from the one presented above because neither the Multiplicity One Theorem nor the index estimates were available at the time. A detailed sketch of the argument can be found in [19].

2.4 Denseness and Equidistribution of Minimal Hypersurfaces

2.4.1 Denseness of Minimal Hypersurfaces

Let M be the set of all smooth metrics with the C^∞-topology. We now present the following result, due to Irie and the authors [13].

Denseness Theorem 2.1.5 *Let M^{n+1} be a closed manifold of dimension $(n + 1)$, with $3 \leq (n + 1) \leq 7$.*

For a C^∞-generic Riemannian metric g on M, the union of all closed, smooth, embedded minimal hypersurfaces is dense.

Proof Given a metric $g \in M$, let $S(g)$ denote the set of all connected, smooth, embedded minimal hypersurfaces with respect to the metric g. An element $\Sigma \in S(g)$ is nondegenerate if every Jacobi vector field vanishes.

Chose an open set $U \subset M$ and set

$$M_U = \{g \in M : \exists \Sigma \in S(g) \text{ with } \Sigma \cap U \neq \emptyset \text{ and } \Sigma \text{ is nondegenerate}\}.$$

The set M_U is open because if $\Sigma \in S(g)$ is nondegenerate, an application of the Inverse Function Theorem implies that for every Riemannian metric g' sufficiently close to g, there exists a unique nondegenerate closed, smooth, embedded minimal hypersurface Σ' close to Σ. In particular, $\Sigma' \cap U \neq \emptyset$ if g' is sufficiently close to g.

If we show that M_U is dense in M then the result follows: Indeed, choose $\{U_i\}_{i \in \mathbb{N}}$ a countable basis of M and consider the set $\cap_i M_{U_i}$, which is C^∞ Baire-generic in M because each M_{U_i} is open and dense in M. Thus if g is a metric in $\cap_i M_{U_i}$ then for every open set $V \subset M$ there is $\Sigma \in S(g)$ that intersects V and this proves the theorem.

Consider the set

$$M_U^* = \{g \in M : \exists \Sigma \in S(g) \text{ with } \Sigma \cap U \neq \emptyset\}.$$

In Proposition 2.3 of [13] it is shown that M_U is dense in M_U^* and so it suffices to see that M_U^* is dense in M.

Let g be an arbitrary smooth Riemannian metric on M and B be an arbitrary neighborhood of g in the C^∞-topology. By the Bumpy Metrics Theorem of White

(Theorem 2.1, [30]), there exists $g' \in \mathcal{B}$ such that every closed, smooth immersed minimal hypersurface with respect to g' is nondegenerate.

Since g' is bumpy, it follows from Sharp (Theorem 2.3 and Remark 2.4, [26]) that the set of connected, closed, smooth, embedded minimal hypersurfaces in (M, g') with both area and index uniformly bounded from above is finite, which means that the set $\mathcal{S}(g')$ is countable and thus

$$C = \{\text{vol}_{g'}(V) : V \text{ a smooth embedded cycle}\}$$

is also countable.

Consider a small perturbation $(g'(t))_{0 \leq t \leq t_0}$ of g' that is supported in U and so that $\text{vol}(M, g'(t_0)) > \text{vol}(M, g')$. For instance, choose $h : M \to \mathbb{R}$ a smooth nonnegative function such that $\text{supp}(h) \subset U$ and $h(x) > 0$ for some $x \in U$, define $g'(t) = (1 + th)g'$ for $t \geq 0$, and let $t_0 > 0$ be sufficiently small so that $g'(t) \in \mathcal{B}$ for every $t \in [0, t_0]$. Because $\text{vol}(M, g'(t_0)) > \text{vol}(M, g')$ it follows from the Weyl Law for the Volume Spectrum 2.3.4 that there exists $k \in \mathbb{N}$ such that $\omega_k(M, g'(t_0)) > \omega_k(M, g')$.

Assume by contradiction $\mathcal{B} \cap \mathcal{M}_U^* = \emptyset$. In this case, for every $t \in [0, t_0]$, every closed, smooth, embedded minimal hypersurface in $(M, g'(t))$ is contained in $M \backslash U$. Since $g'(t) = g'$ outside a compact set contained in U we have $\mathcal{S}(g') = \mathcal{S}(g'(t))$ and so we conclude from Theorem 2.3.10 that $\omega_k(M, g'(t)) \in C$ for all $t \in [0, t_0]$. But C is countable and we know from Proposition 2.3.3 that the function $t \mapsto \omega_k(M, g'(t))$ is continuous. Hence $t \mapsto \omega_k(M, g'(t))$ is constant in the interval $[0, t_0]$. This contradicts the fact that $\omega_k(M, g'(t_0)) > \omega_k(M, g')$.

Therefore $\mathcal{B} \cap \mathcal{M}_U^* \neq \emptyset$ and hence $\mathcal{M}_U^*$ is dense in $\mathcal{M}$.   □

2.4.2 Equidistribution of Minimal Hypersurfaces

In this section we explain the key ideas behind the following result.

Equidistribution Theorem 2.1.6 *Let M^{n+1} be a closed manifold of dimension $(n + 1)$, with $3 \leq (n + 1) \leq 7$.*

For a C^∞-generic Riemannian metric g on M, there exists a sequence $\{\Sigma_j\}_{j \in \mathbb{N}}$ of closed, smooth, embedded, connected minimal hypersurfaces that is equidistributed in M: for any $f \in C^0(M)$ we have

$$\lim_{q \to \infty} \frac{1}{\sum_{j=1}^q \text{vol}_g(\Sigma_j)} \sum_{j=1}^q \int_{\Sigma_j} f \, d\Sigma_j = \frac{1}{\text{vol}_g M} \int_M f \, dV.$$

Before we sketch its proof we discuss an heuristic argument. Fix $g \in \mathcal{M}$, choose $f \in C^\infty(M)$ and a small closed interval $I \subset \mathbb{R}$ containing the origin in its interior. For each $t \in I$ define $g(t) = \exp(tf)g$ and, for each $k \in \mathbb{N}$, consider the function

$$t \mapsto \mathcal{W}_k(t) = \ln \omega_k(M, g(t)) - \frac{n}{n+1} \ln \mathrm{vol}_{g(t)}(M) - \ln k^{\frac{n}{n+1}}. \qquad (2.4.1)$$

From Proposition 2.3.3 we have that $\mathcal{W}_k$ is uniformly Lipschitz on I (independently of $k \in \mathbb{N}$). Thus the Weyl Law for the Volume Spectrum 2.3.4 implies that

$$\lim_{k \to \infty} \max\{\mathcal{W}_k(t) - \ln a(n) : t \in I\} = 0. \qquad (2.4.2)$$

Recall the definition of minimal embedded cycles in Definition 2.2.1. We now assume the following strong assumption: For all $k \in \mathbb{N}$ and $t \in I$ there is a *unique* smooth embedded minimal cycle $\Sigma_k(t)$ (with respect to $g(t)$) so that

- $\mathrm{vol}_{g(t)}(\Sigma_k(t)) = \omega_k(M, g(t))$;
- $\Sigma_k(t)$ is two-sided and multiplicity one.

The uniqueness of $\Sigma_k(t)$ and the Sharp Compactness Theorem [26] implies that, for all $k \in \mathbb{N}$, the deformation $t \in I \mapsto \Sigma_k(t)$ is smooth and so, using the fact that $\partial_t g(t) = fg(t)$, we have

$$\frac{d}{dt}\omega_k(M, g(t)) = \frac{d}{dt}\mathrm{vol}_{g(t)}(\Sigma_k(t)) = \frac{n}{2} \int_{\Sigma_k(t)} f \, d\Sigma_k(t)$$

$$\frac{d}{dt}\mathrm{vol}_{g(t)}(M) = \frac{n+1}{2} \int_M f \, dV_{g(t)}.$$

Hence, we have that for all $t \in I$

$$\mathcal{W}_k'(t) = \frac{n}{2} \left(\frac{1}{\mathrm{vol}_{g(t)}(\Sigma_k(t))} \int_{\Sigma_k(t)} f \, d\Sigma_k(t) - \frac{1}{\mathrm{vol}(M)} \int_M f \, dV_{g(t)} \right).$$

Thus we deduce from (2.4.2) that, after passing to a subsequence, $\mathcal{W}_{k_j}'(t) \to 0$ for almost all $t \in I$. Hence setting $\Sigma^j = \Sigma_{k_j}(t)$ we deduce

$$\lim_{j \to \infty} \frac{1}{\mathrm{vol}_g(\Sigma^j)} \int_{\Sigma^j} f \, d\Sigma^j = \frac{1}{\mathrm{vol}(M)} \int_M f \, dV.$$

Without assuming the strong assumption above, the function $\mathcal{W}_k$ is only uniformly Lipschitz and one can surely find a sequence of uniformly Lipschitz functions converging to a constant whose derivative (whenever it is well defined) is uniformly away from zero. Overcoming the lack of differentiability everywhere of $\mathcal{W}_k$ is at the core of the proof of the Equidistribution Theorem 2.1.6.

Given a Lipschitz function ϕ on a cube $I^m \subset \mathbb{R}^m$ we know from Rademacher Theorem that ϕ is differentiable almost everywhere. Let us consider the *generalized derivative* of ϕ that is defined as

$$\partial^*\phi(t) = \text{Conv}\{\lim_{i\to\infty} \nabla\phi(t_i) : \nabla\phi(t_i) \text{ exists and } \lim t_i = t\},$$

where $\text{Conv}(K)$ denotes the convex hull of $K \subset \mathbb{R}^m$. If the function is C^1, the generalized derivative coincides with the classical derivative. The result we need is the following.

Lemma 2.4.1 *There is a constant C (depending on I^m) so that for every Lipschitz function ϕ on I^m with*

$$|\phi(x) - \phi(y)| \le \varepsilon \quad \text{for all } x, y \in I^m,$$

there is $\bar{t} \in I^m$ with $\text{dist}(0, \partial^*\phi(\bar{t})) \le C\varepsilon$.

Proof Let us first assume that ϕ achieves its maximum at an interior point $\bar{t} \in I^m$. We now argue that $0 \in \partial^*\phi(\bar{t})$.

Consider the compact set

$$K = \{\lim_{i\to\infty} \nabla\phi(t_i) : \nabla\phi(t_i) \text{ exists and } \lim t_i = \bar{t}\}.$$

For almost all unit vector w in $\mathbb{R}^m$ we have that the function $\phi_w(s) = \phi(\bar{t} + sw)$ is defined in an open neighborhood U of zero and $\nabla\phi(\bar{t} + sw)$ is well defined for almost all $s \in U$. The function ϕ_w has an absolute maximum at $s = 0$, its derivative is given by $\phi'_w(s) = \nabla\phi(\bar{t} + sw).w$ for s almost everywhere, and thus there is $v \in K$ with $v.w \le 0$. Hence we deduce that for every unit vector $w \in \mathbb{R}^m$ there is $v \in K$ with $v.w \ge 0$. If $0 \notin \text{Conv}(K)$ there is $0 \ne p \in \text{Conv}(K)$ that minimizes the distance to the origin and so $v.p \ge |p|^2$ for all $v \in \text{Conv}(K)$, which is a contradiction.

We now handle the general case. Choose a smooth function $\eta : I^m \to \mathbb{R}$ with $\eta(0) = 0$ and $\eta = 2$ on ∂I^m. Set $\psi = \phi - \varepsilon\eta$. Then $\psi(x) \le \psi(0) - \varepsilon$ for all $x \in \partial I^m$ and so ψ must have an interior maximum $\bar{t}$. Thus $0 \in \partial^*\psi(\bar{t})$ and so $\text{dist}(0, \partial^*\phi(\bar{t})) \le C\varepsilon$, where C bounds the gradient of η. $\qquad\square$

Sketch of Proof of Theorem 2.1.6 Given a metric $g \in \mathcal{M}$, $\mathcal{S}(g)$ denotes the set of all connected, smooth, embedded minimal hypersurfaces with respect to g and $\mathcal{V}(g)$ denotes the set of all smooth embedded minimal cycles. Given $S \in \mathcal{V}(g)$ we define by μ_S and μ_M the unit Radon measures on M given by, respectively,

$$\mu_S(f) = \frac{\|S\|(f)}{\|S\|(M)} \text{ and } \mu_M(f) = \frac{1}{\text{vol}_g(M)} \int_M f \, dV, \quad f \in C^0(M).$$

Finally, we define by $\text{Conv}(\mathcal{V}(g))$ the unit Radon measures μ that are given by convex linear combinations of Radon measures μ_S, i.e.,

$$\left\{ \sum_{i=1}^{J} a_i \mu_{S_i} : 0 \leq a_i \leq 1, S_i \in \mathcal{V}(g), i = 1, \ldots, J \text{ and } a_1 + \ldots + a_J = 1 \right\}.$$

We say that $\mu = \sum_{i=1}^{J} a_i \mu_{S_i} \in \text{Conv}(\mathcal{V}(g))$ is non-degenerate if the support of each S_i is a non-degenerate minimal hypersurface.

Choose a subset $\{\psi_i\}_{i \in \mathbb{N}} \subset C^\infty(M)$ that is dense in $C^0(M)$ and set*

$$M(m) = \{g \in M : \exists \mu \in \text{Conv}(\mathcal{V}(g)) \text{ non-degenerate such that}$$

$$|\mu(\psi_i) - \mu_M(\psi_i)| < m^{-1} \quad \text{for all } i = 1, \ldots, m\}.$$

A standard perturbation argument based on the Inverse Function Theorem shows that $M(m)$ is open in the C^∞-topology for all $m \in \mathbb{N}$.

We now explain that if $M(m)$ is dense in M for all $m \in \mathbb{N}$ then the desired result follows. If so $M_\infty = \cap_{m \in \mathbb{N}} M(m)$ is a residual set (in the Baire sense) and we choose $g \in M_\infty$. In this case we can find a sequence $\{\mu_m\}_{m \in \mathbb{N}}$ of elements of $\text{Conv}(\mathcal{V}(g))$ so that

$$|\mu_m(\psi_i) - \mu_M(\psi_i)| < m^{-1} \quad \text{for all } i = 1, \ldots, m.$$

Hence every accumulation point ν (in the weak topology) is a Radon measure with $\nu(\psi_i) = \mu_M(\psi_i)$ for all $i \in \mathbb{N}$. Thus, from the way $\{\psi_i\}_{i \in \mathbb{N}}$ was chosen, the sequence $\{\mu_m\}_{m \in \mathbb{N}}$ converges weakly to μ_M.

We have for some $J_m \in \mathbb{N}$

$$\mu_m = \sum_{j=1}^{J_m} a_{j,m} \mu_{S_{j,m}}, 0 \leq a_{j,m} \leq 1, S_{j,m} \in \mathcal{V}(g) \text{ for all } j = 0, \ldots, J_m$$

and $\sum_{j=1}^{J_m} a_{j,m} = 1$. Choose integers $d_m, b_{j,m}, j = 1, \ldots, J_m$ so that

$$\left| \frac{a_{j,m}}{||S_{j,m}||(M)} - \frac{b_{j,m}}{d_m} \right| < \frac{1}{m J_m ||S_{j,m}||(M)} \tag{2.4.3}$$

and set $V_{j,m} = b_{j,m} S_{j,m} \in \mathcal{V}(g)$. We claim that for all $f \in C^0(M)$ we have

$$\lim_{m \to \infty} \frac{\sum_{j=1}^{J_m} ||V_{j,m}||(f)}{\sum_{j=1}^{J_m} ||V_{j,m}||(M)} = \mu_M(f). \tag{2.4.4}$$

With $f \in C^0(M)$ fixed and $K = \sup_M |f|$ we have, using (2.4.3),

$$\mu_m(f) = \sum_{j=1}^{J_m} a_{j,m} \mu_{S_{j,m}}(f) = \sum_{j=1}^{J_m} \frac{b_{j,m}}{d_m} \|S_{j,m}\|(f) + \sum_{j=1}^{J_m} O\left(\frac{K}{m J_m}\right)$$

$$= \sum_{j=1}^{J_m} \frac{b_{j,m}}{d_m} \|S_{j,m}\|(f) + O\left(\frac{K}{m}\right) = \frac{\sum_{j=1}^{J_m} \|V_{j,m}\|(f)}{d_m} + O\left(\frac{K}{m}\right).$$

Furthermore, combining $\sum_{j=1}^{J_m} a_{j,m} = 1$ with (2.4.3) we have

$$\frac{\sum_{j=1}^{J_m} \|V_{j,m}\|(M)}{d_m} = \frac{\sum_{j=1}^{J_m} b_{j,m}\|S_{j,m}\|(M)}{d_m} = 1 + O\left(\frac{1}{m}\right),$$

which when combined with the previous identities gives

$$\mu_m(f) = \left(1 + O\left(\frac{1}{m}\right)\right) \frac{\sum_{j=1}^{J_m} \|V_{j,m}\|(f)}{\sum_{j=1}^{J_m} \|V_{j,m}\|(M)} + O\left(\frac{K}{m}\right).$$

Making $m \to \infty$ we deduce (2.4.4). One immediate consequence is that we obtain the existence of a finite sequence $\{\Sigma_{i,m}\}_{i=1}^{P_m}$ of elements in $\mathcal{S}(g)$ so that for all $f \in C^0(M)$ we have

$$\lim_{m \to \infty} \frac{1}{\sum_{i=1}^{P_m} \mathrm{vol}_g(\Sigma_{i,m})} \sum_{i=1}^{P_m} \int_{\Sigma_{i,m}} f \, d\Sigma_{i,m} = \frac{1}{\mathrm{vol}_g(M)} \int_M f \, dV.$$

Using this identity, a further combinatorial argument (see [22, pages 15, 16]) shows that we can extract a sequence $\{\Sigma_i\}_{i \in \mathbb{N}}$ of elements of $\cup_{m \in \mathbb{N}} \{\Sigma_{j,m}\}_{j=1}^{P_m}$ so that for all $f \in C^0(M)$

$$\lim_{q \to \infty} \frac{1}{\sum_{j=1}^{q} \mathrm{vol}_g(\Sigma_j)} \sum_{j=1}^{q} \int_{\Sigma_j} f \, d\Sigma_j = \frac{1}{\mathrm{vol}_g M} \int_M f \, dV.$$

We now show that $\mathcal{M}(m)$ is dense in $\mathcal{M}$ with respect to the C^∞-topology. Consider the slightly larger set

$$\mathcal{M}^*(m) = \{g \in \mathcal{M} : \exists \mu \in \mathrm{Conv}(\mathcal{V}(g)) \text{ such that}$$

$$|\mu(\psi_i) - \mu_M(\psi_i)| < m^{-1} \quad \text{for all } i = 1, \dots, m\}.$$

The first remark is that using Lemma 4 of [22] one can see that $M^*(m)$ and $M(m)$ have the same closure in M (the reader can see the details at the end of Section 3 of [22]). Thus it suffices to see that $M^*(m)$ is dense in M.

Choose $I^m \subset \mathbb{R}^m$ a small cube centered at the origin. For every vector $t \in I^m$ define $g(t) = \exp(\sum_{i=1}^m t_i \psi_i) g$ and, for each $k \in \mathbb{N}$, consider the functions $\mathcal{W}_k(t)$ defined in (2.4.1). Like it was explained during the heuristic argument, we have from Proposition 2.3.3 that $\mathcal{W}_k$ is uniformly Lipschitz on I^m (independently of $k \in \mathbb{N}$) and the Weyl Law for the Volume Spectrum 2.3.4 implies that

$$\lim_{k \to \infty} \max\{\mathcal{W}_k(t) - \ln a(n) : t \in I^m\} = 0. \tag{2.4.5}$$

From Lemma 2 of [22] (which is based on Smale's Transversality Theorem) there are arbitrarily small smooth perturbations of the map $t \mapsto g(t)$ that are bumpy for almost all $t \in I^m$. We leave the details for the reader to see in [22] and instead assume, for simplicity, that $g(t)$ is a bumpy metric for almost all $t \in I^m$. Using the fact that $\partial_{t_i} g(t) = \psi_i g(t)$, it is shown in [22, Lemma 2] that for a full measure set $t \in A \subset I^m$ we have

$$\frac{\partial}{\partial t_i} \omega_k(M, g(t)) = \frac{n}{2} ||S_k(t)||(\psi_i), \quad i = 1, \ldots, m, \tag{2.4.6}$$

where $S_k(t) \in \mathcal{V}(g(t))$ is a smooth embedded minimal cycle with

$$\omega_k(M, g(t)) = ||S_k(t)||(M) \quad \text{and} \quad \text{index}(S_k(t)) \leq k.$$

Combining with the fact that

$$\frac{\partial}{\partial t_i} \text{vol}_{g(t)}(M) = \frac{n+1}{2} \int_M \psi_i \, dV_{g(t)}$$

we have that for almost all $t \in I^m$

$$\frac{\partial}{\partial t_i} \mathcal{W}_k(t) = \frac{n}{2} \left(\mu_{S_k(t)}(\psi_i) - \mu_M(\psi_i) \right) \quad i = 1, \ldots, m. \tag{2.4.7}$$

The sequence of functions $\{\mathcal{W}_k\}_{k \in \mathbb{N}}$ converges uniformly to a constant (2.4.5) as $k \to \infty$ and so we deduce from Lemma 2.4.1 the existence of $k \in \mathbb{N}$ and $\bar{t} \in I^m$ so that $\text{dist}(0, \partial^* \mathcal{W}_k(\bar{t})) < n/(2m)$.

Choose $v \in \partial^* \mathcal{W}_k(\bar{t})$ with $|v| < n/(2m)$ and set

$$K = \{ \lim_{i \to \infty} \nabla \mathcal{W}_k(t_i) : \nabla \mathcal{W}_k(t_i) \text{ exists and } \lim t_i = \bar{t} \} \subset \mathbb{R}^m.$$

From Caratheodory Theorem, there are $v_0, \ldots, v_m \in K$ and $0 \leq a_0, \ldots, a_m \leq 1$ so that

$$v = \sum_{l=0}^{m} a_l v_l \quad \text{and} \quad a_0 + \ldots + a_l = 1.$$

Claim For each $l = 0, \ldots, m$ there is a smooth embedded cycle V_l so that

$$v_l.e_i = \frac{n}{2} \left(\mu_{V_l}(\psi_i) - \mu_M(\psi_i) \right) \quad i = 1, \ldots, m,$$

where e_i is the ith coordinate vector.

There is a sequence $\{t_j\}_{j \in \mathbb{N}} \in A \subset I^m$ so that $t_j \to \bar{t}$ and $\nabla \mathcal{W}_k(t_j) \to v_l$ as $t \to \infty$. Considering the smooth embedded minimal cycles $S_k(t_j)$ that appear in (2.4.7), we have from Sharp Compactness Theorem [26, Theorem 2.3] the existence of a smooth embedded minimal cycle V_l so that, after passing to a subsequence, $S_k(t_j)$ converges to V_l in the varifold sense. Thus $||S_k(t_j)||(M) \to ||V_l||(M)$ and $||S_k(t_j)||(\psi_i) \to ||V_l||(\psi_i)$, $i = 1, \ldots, m$, as $j \to \infty$. Therefore we have from (2.4.7) that

$$v_l.e_i = \lim_{j \to \infty} \nabla \mathcal{W}_k(t_j).e_i = \lim_{j \to \infty} \frac{\partial}{\partial t_i} W_k(t_j)$$

$$= \lim_{j \to \infty} \frac{n}{2} \left(\mu_{S_k(t_j)}(\psi_i) - \mu_M(\psi_i) \right) = \frac{n}{2} \left(\mu_{V_l}(\psi_i) - \mu_M(\psi_i) \right)$$

for all $i = 1, \ldots, m$.

The claim implies that if we consider $\mu = \sum_{l=0}^{m} a_l \mu_{V_l} \in \text{Conv}(\mathcal{V}(g(\bar{t})))$ then

$$v.e_i = \sum_{l=0}^{m} a_l v_l.e_i = \sum_{l=0}^{m} a_l \frac{n}{2} \left(\mu_{V_l}(\psi_i) - \mu_M(\psi_i) \right) = \frac{n}{2} \left(\mu(\psi_i) - \mu_M(\psi_i) \right)$$

for all $i = 1, \ldots, m$. Recalling that $|v| < n/(2m)$, the identity above implies that $g(\bar{t}) \in \mathcal{M}^*(m)$. Because the cube $I^m \subset \mathbb{R}^m$ can be chosen arbitrarily small, we deduce that $\mathcal{M}^*(m)$ is dense in $\mathcal{M}$. $\square$

References

1. F. Almgren, The homotopy groups of the integral cycle groups. Topology **1**(4), 257–299 (1962)
2. F. Almgren, in *The Theory of Varifolds*. Mimeographed Notes (Princeton, 1965)
3. T. Beck, S. Becker-Kahn, B. Hanin, Nodal sets of smooth functions with finite vanishing order and p-sweepouts. Calc. Var. Partial Differ. Equ. **57**(5), 140 (2018)

4. V. Buchstaber, T. Panov, in *Torus Actions and Their Applications in Topology and Combinatorics*. University Lecture Series, vol. 24 (American Mathematical Society, Providence, 2002), viii+144 pp

5. O. Chodosh, C. Mantoulidis, Minimal surfaces and the Allen–Cahn equation on 3-manifolds: index, multiplicity, and curvature estimates. Ann. Math. **191**(1), 213–328 (2020). arXiv:1803.02716 [math.DG]

6. H. Donnelly, C. Fefferman, Nodal sets of eigenfunctions on Riemannian manifolds. Invent. Math. **93**, 161–183 (1988)

7. H. Federer, in *Geometric Measure Theory*. Die Grundlehren der mathematischen Wissenschaften, Band 153 (Springer, New York 1969)

8. W.H. Fleming, Flat chains over a finite coefficient group. Trans. Am. Math. Soc. **121**, 160–186 (1966)

9. M. Gromov, Dimension, nonlinear spectra and width, in *Geometric Aspects of Functional Analysis (1986/1987)*. Lecture Notes in Mathematics, vol. 1317 (Springer, Berlin, 1988), pp. 132–184

10. M. Gromov, Isoperimetry of waists and concentration of maps. Geom. Funct. Anal. **13**, 178–215 (2003)

11. L. Guth, Minimax problems related to cup powers and Steenrod squares. Geom. Funct. Anal. 18, 1917–1987 (2009)

12. A. Hatcher, *Algebraic Topology* (Cambridge University Press, Cambridge, 2002) arXiv:1709.02652 (2017)

13. K. Irie, F.C. Marques, A. Neves, Density of minimal hypersurfaces for generic metrics. Ann. Math. **187**(3), 963–972 (2018)

14. D. Jerison, G. Lebeau, Nodal sets of sums of eigenfunctions, in *Harmonic Analysis and Partial Differential Equations: Essays in Honour of Alberto P. Calderón*, ed. by M. Christ, C. Kenig, C. Sadosky (University of Chicago Press, Chicago, 1999), pp. 223–239

15. B. Lawson, Complete minimal surfaces in S^3. Ann. Math. **92** , 335–374 (1970)

16. Y. Liokumovich, F.C. Marques, A. Neves, Weyl law for the volume spectrum. Ann. Math. **187**(3), 933–961 (2018)

17. F.C. Marques, A. Neves, Min–max theory and the Willmore conjecture. Ann. Math. **179**(2), 683–782 (2014)

18. F.C. Marques, A. Neves, Existence of infinitely many minimal hypersurfaces in positive Ricci curvature. Invent. Math. **209**, 577–616 (2017)

19. F.C. Marques, A. Neves, in *Applications of Almgren–Pitts Min–Max Theory*. Current Developments in Mathematics, vol. 2013 (International Press, Somerville, 2014), pp. 1–71

20. F.C. Marques, A. Neves, Morse index and multiplicity of min–max minimal hypersurfaces. Camb. J. Math. **4**(4), 463–511 (2016)

21. F.C. Marques, A. Neves, Morse index of multiplicity one min–max minimal hypersurfaces (2018). arXiv:1803.04273 [math.DG]

22. F.C. Marques, A. Neves, A. Song, Equidistribution of minimal hypersurfaces for generic metrics. Invent. Math. **216**, 421–443 (2019). arXiv:1712.06238

23. C. Nurser, Low min–max widths of the round three-sphere. Ph.D. thesis, 2016

24. J. Pitts, in *Existence and Regularity of Minimal Surfaces on Riemannian Manifolds*. Mathematical Notes, vol. 27 (Princeton University Press, Princeton, 1981)

25. R. Schoen, L. Simon, Regularity of stable minimal hypersurfaces. Commun. Pure Appl. Math. **34**, 741–797 (1981)

26. B. Sharp, Compactness of minimal hypersurfaces with bounded index. J. Differ. Geom. **106**(2), 317–339 (2017)

27. L. Simon, Lectures on geometric measure theory, in *Proceedings of the Centre for Mathematical Analysis* (Australian National University, Canberra, 1983)

28. A. Song, Existence of infinitely many minimal hypersurfaces in closed manifolds (2018). arXiv:1806.08816 [math.DG]

29. H. Weyl, in *Über die Asymptotische Verteilung der Eigenwerte*. Nachr. Konigl. Ges. Wiss (Göttingen, 1911), pp. 110–117

30. B. White, The space of minimal submanifolds for varying Riemannian metrics. Indiana Univ. Math. J. **40**, 161–200 (1991)
31. B. White, On the bumpy metrics theorem for minimal submanifolds. Am. J. Math. **139**(4), 1149–1155 (2017)
32. S.-T. Yau, Problem section, in *Seminar on Differential Geometry*. Annals of Mathematics Studies, vol. 102 (Princeton University Press, Princeton, 1982), pp. 669–706
33. W. Ziemer, Integral currents mod 2. Trans. Am. Math. Soc. **105**, 496–524 (1962)
34. X. Zhou, J. Zhu, Min–max theory for constant mean curvature hypersurfaces. Invent. Math. **218**, 441–490 (2019). arXiv:1707.08012 [math.DG]
35. X. Zhou, On the multiplicity one conjecture in min–max theory (2019). arXiv:1901.01173 [math.DG]
36. X. Zhou, J. Zhu, Existence of hypersurfaces with prescribed mean curvature I—generic min–max (2018). arXiv:1808.03527 [math.DG]

Chapter 3
Ricci Flow and Ricci Limit Spaces

Peter M. Topping

Abstract Ricci flow gives us a natural way of evolving a Riemannian manifold in order to even out the geometry. It has been fundamental in the resolution of a number of famous open problems in geometry and topology such as the Poincaré conjecture and Thurston's Geometrisation conjecture. In this chapter Ricci flow theory is studied in unfamiliar situations in order to solve different types of problems. Everything revolves around understanding flows with unbounded curvature on noncompact manifolds, possibly with very rough initial data, and the lectures describe some of the new phenomena that arise in these situations. After describing a complete theory for two-dimensional underlying manifolds, recent work is described in the three-dimensional case with applications to Ricci limit spaces.

3.1   Introduction

In these lecture notes I hope to describe some of the developments in Ricci flow theory from the past decade or so in a manner accessible to starting graduate students. I will be most concerned with the understanding of Ricci flows that are permitted to have unbounded curvature in the sense that the curvature can blow up as we wander off to spatial infinity and/or as we decrease time to some singular time. This is an area that has seen substantial recent progress, and it has long been understood that a sufficiently developed theory will be useful in settling important open problems in geometry, particularly to the subject of understanding the topology of manifolds that satisfy certain curvature restrictions. In these notes I will outline an application of the theory as developed so far, but instead to the theory of so-called Ricci limit spaces, as developed particularly by Cheeger–Colding around 20 years ago.

P. M. Topping (✉)
Mathematics Institute, University of Warwick, Coventry, UK
e-mail: p.m.topping@warwick.ac.uk

M. J. Gursky, A. Malchiodi (eds.), *Geometric Analysis*, Lecture Notes
in Mathematics 2263, https://doi.org/10.1007/978-3-030-53725-8_3

Much of the work I describe is joint work with Miles Simon. Some of the general ideas will be explained in the context of two-dimensional flows—the case in which these were developed—which is partly joint work with Gregor Giesen.

These notes have been written for a summer school at Cetraro in 2018 and I would like to thank Matt Gursky and Andrea Malchiodi for the invitation to explain this theory. A subset of the lectures were given at the Max Planck Institute for Mathematics in Bonn in 2017, and I would like to thank Karl-Theodor Sturm for that invitation.

3.2 Introductory Material

3.2.1 Ricci Flow Basics

Traditionally, the Ricci flow has been posed as follows. Given a Riemannian manifold (M, g_0), we ask whether we can find a smooth one-parameter family of Riemannian metrics $g(t)$ for $t \in [0, T)$ such that $g(0) = g_0$ and satisfying the PDE

$$\frac{\partial g}{\partial t} = -2\mathrm{Ric}_g, \tag{3.1}$$

where Ric_g is the Ricci curvature of g. If you are new to this subject then you should view the right-hand side of (3.1) morally as a type of Laplacian of g, and the equation as an essentially parabolic PDE for the metric g. One of the simplest examples that is not completely trivial is the shrinking sphere: if (M, g_0) is the round unit sphere of dimension n, then one can easily check that

$$g(t) = \big(1 - 2(n - 1)t\big)g_0 \tag{3.2}$$

is a solution for $t \in [0, \frac{1}{2(n-1)})$. Another standard example that is important to digest is the dumbbell example: If we connect two round three-spheres by a thin neck $\varepsilon S^2 \times [0, 1]$ to give a nonround three-sphere, then Ricci flow will deform it by shrinking the thin neck, which will degenerate before much happens to the two original three-spheres. See [30, Section 1.3.2] for a more involved discussion and pictures.

The Ricci flow is a little simpler if the evolving manifold is Kähler, and the simplest of all is in the lowest possible dimension, i.e. $\dim M = 2$, which corresponds to one complex-dimensional *Kähler Ricci flow*. Then we can write $\mathrm{Ric}_g = Kg$, where K is the Gauss curvature, and the flow can be seen to deform the metric conformally.

The Ricci flow was introduced by Hamilton in 1982 [16] in order to prove the following theorem.[1]

Theorem 3.1 (Hamilton [16]) *If (M, g_0) is a closed, simply connected Riemannian 3-manifold with $\mathrm{Ric} > 0$, then it is diffeomorphic to S^3.*

The rough strategy is as follows. First, take (M, g_0) and run the Ricci flow. Second, show that the condition $\mathrm{Ric} > 0$ is preserved for each of the subsequent metrics $g(t)$. (It is worth remarking that a local form of this preservation of lower Ricci control has recently been proved [28] and will be pivotal later on; see Lemma 3.5.) Finally, show that after appropriate scaling, $(M, g(t))$ converges to a manifold of constant sectional curvature as time increases. This must then be S^3.

For more details of this argument, from a modern point of view, and a more involved introduction to Ricci flow, see [30].

This application is a model for many of the most remarkable subsequent applications of Ricci flow. That is, if we assume that a manifold satisfies a certain curvature condition, then often we can use Ricci flow to deform it smoothly to something whose topology we can identify (or to a *collection* of objects whose topology we can identify). The famous work of Perelman goes one step further, in some sense, by carrying out that programme in three dimensions without any curvature hypothesis whatsoever, thus proving the Poincaré conjecture and Thurston's geometrisation conjecture as originally envisaged by Hamilton and Yau.

In contrast, in these lectures we will use Ricci flow for a completely different task. We will have to pose Ricci flow in a different way, specifying initial data that is a lot rougher than a Riemannian metric, and as a result we will get a refined description of so-called Ricci limit spaces as we explain later. However, the developments in the Ricci flow theory that permit these new applications seem also likely to lead to new applications of the traditional type, i.e. assume a curvature condition and deduce a topological conclusion.

3.2.2 Traditional Theory: Closed or Bounded Curvature

The main difference between the theory we will see in these lectures and the traditional theory of Ricci flow is that most traditional applications concern *closed* manifolds, whereas we are interested in general noncompact manifolds. In the closed case, Hamilton proved existence of a unique solution to Ricci flow, as posed above. In the non-closed case, things become mysterious and involved, unless one reduces essentially to a situation like the closed case by imposing a constraint of *bounded curvature*.

---

[1]When we write $\mathrm{Ric} > 0$, we mean that $\mathrm{Ric}(X, X) > 0$ for every nonzero tangent vector X.

Theorem 3.2 (Shi [26] and Hamilton [16]) *Suppose (M, g_0) is a complete* **bounded curvature** *Riemannian manifold of arbitrary dimension—let's say* $|\mathrm{Rm}|_{g_0} \leq K$ *for some* $K > 0$. *Then there exists a Ricci flow* $g(t)$ *on* M *for* $t \in [0, \frac{1}{16K}]$ *with* $g(0) = g_0$ *such that* $|\mathrm{Rm}|_{g(t)} \leq 2K$ *for each* t.

Here we are writing Rm for the full curvature tensor. For fixed dimension, controlling $|\mathrm{Rm}|$ is a little like controlling the magnitude of the sectional curvature.

Because the Ricci flow of Theorem 3.2 has bounded curvature, it will automatically inherit the completeness of g_0, i.e. at each time t, the metric $g(t)$ will be complete. This is because the lengths of paths can only grow or shrink exponentially:

Lemma 3.1 *Given a Ricci flow* $g(t)$, $t \in [0, T]$, *with* $-M_1 \leq \mathrm{Ric}_{g(t)} \leq M_2$, *we have*

$$e^{-2M_2 t} g(0) \leq g(t) \leq e^{2M_1 t} g(0),$$

for each t.

Here, when we write $\mathrm{Ric}_g \leq K$, say, we mean that $\mathrm{Ric}_g \leq Kg$ as bilinear forms.

Thus, distances can only expand by a factor at most $e^{M_1 t}$, and shrink by a factor no more extreme than $e^{-M_2 t}$ after a time t. This lemma is a simple computation. For example, for a fixed tangent vector X, we see that

$$\frac{\partial}{\partial t} \ln g(X, X) \leq 2M_1.$$

We also remark that, as proved by Chen and Zhu [6] and simplified by Kotschwar [21], the flow of Theorem 3.2 is unique in the sense that two smooth bounded-curvature Ricci flows starting at the same smooth, complete, bounded curvature initial metric must agree. We emphasise that by *bounded-curvature* Ricci flow we mean that there is a t-independent curvature bound, not that the curvature of each $g(t)$ is bounded, otherwise the uniqueness fails (see, e.g., Example 3.1 below).

Remark 3.1 For how long does Shi's flow live? The theorem as we gave it tells us that we can at least flow for a time $1/(16K)$, which is, by chance, independent of the dimension of M. But the theorem can be iterated to give a flow that lives until the curvature blows up, and because the theorem provides a curvature bound at the final time, the iterated flow must live for at least a time $1/(16K) + 1/(32K) + 1/(64K) + \cdots = 1/(8K)$. None of these explicit times are claimed to be optimal. Traditionally one called a flow whose curvature blew up at a finite end time a *maximal* flow. However, it was shown in [2] and [14] that one can have flows in this context whose curvature blows up only at spatial infinity at the end time, and which can be extended beyond to exist for all time (see also Sect. 3.3.3).

3.3 Noncompact and Unbounded Curvature Case

3.3.1 Difficulties of the Noncompact Case; Examples to Reboot Intuition

Shi's theorem 3.2 gives a fine existence statement in the special case that we wish to flow a manifold with bounded curvature. Unfortunately, this is an unreasonable assumption in virtually all applications when we are working on a noncompact manifold. Thus we would like to consider the problem of starting a Ricci flow with a more general manifold, particularly one of unbounded curvature, or more generally considering flows that themselves have unbounded curvature. Unfortunately, such flows can be very weird. They can jump back and forth between being complete and being incomplete. They can exhibit unusual nonuniqueness properties. They tend to render the ever-useful maximum principle void. They will try to jump to a different underlying manifold if given the chance. Ultimately, they will simply fail to exist in certain cases.

Let us elaborate on some of these vague statements with some specific examples.

Example 3.1 If (M, g_0) is the two-dimensional hyperbolic plane, then there is a simple expanding solution $g(t) = (1 + 2t)g_0$ for $t \in [0, \infty)$ that is the direct analogue of the shrinking sphere that we saw in (3.2). This is the solution that would be given by Shi's theorem 3.2. However, more surprisingly, there exist infinitely many other smooth Ricci flows also starting with the hyperbolic plane, and with bounded curvature for $t \in (\varepsilon, \infty)$, for any $\varepsilon > 0$. As we will see, these solutions are necessarily not complete, and thus must have unbounded curvature as $t \downarrow 0$, although being smooth they trivially have bounded curvature as $t \downarrow 0$ over any compact subset of the domain. Intuitively they instantly peel away at the boundary, though not necessarily in a rotationally symmetric way. We'll construct them in a moment.

Example 3.2 There exists a rotationally symmetric, complete, bounded curvature conformal metric g_0 on $\mathbb{R}^2$, and a smooth rotationally symmetric Ricci flow $g(t)$ for $t \in [0, 1)$ with $g(0) = g_0$, such that the distance from the origin to infinity (i.e. the length of a radial ray) behaves like $-\log t$ (i.e. it is bounded from above and below by $-\log t$ times appropriate positive constants). In particular, the boundary at infinity is uniformly sucked in making it instantaneously incomplete. The curvature at infinity behaves like t^{-2} (again controlled, up to a constant factor, from above and below).

Before continuing with our examples, let's construct the two we have already seen. They both have the same building block, which is the concept of a Ricci flow that contracts a hyperbolic cusp. We showed in [32] (though the best way of seeing this now is to use the sharp L^1-L^∞ smoothing estimate proved in [34]) that if we are given a surface with a hyperbolic cusp, then one way of flowing it is to add a point at infinity and let it contract.

To see a precise instance of this, consider the punctured two-dimensional disc $D\backslash$ $\{0\}$, equipped with the unique (conformal) complete hyperbolic metric g_h. That is,

$$g_h = \frac{1}{r^2 \log r^2}(dx^2 + dy^2),$$

where $r^2 = x^2 + y^2$. This has a hyperbolic cusp at the origin. Similarly to what we have seen before, there is an explicit Ricci flow starting with g_h given by $g(t) = (1 + 2t)g_h$, and this is the flow that Shi's theorem 3.2 would give. However, in [35] we constructed a smooth Ricci flow $g(t)$ on the whole disc D, for $t \in (0, \infty)$ such that $g(t) \to g_h$ smoothly locally on $D \setminus \{0\}$ as $t \downarrow 0$. This flow we can restrict to $D \setminus \{0\}$, giving a smooth Ricci flow on $D \setminus \{0\}$ for $t \in [0, \infty)$ such that $g(0) = g_h$, but so that $g(t)$ is incomplete at the origin for $t > 0$.

To construct Example 3.1, we merely have to lift this flow on $D \setminus \{0\}$ to its universal cover D.

Meanwhile, to construct Example 3.2, we take a slight variant of the $D \setminus \{0\}$ example above by taking a complete rotationally symmetric metric g_0 on a punctured 2-sphere, with a hyperbolic cusp. Again, we can find a rotationally symmetric Ricci flow on the whole 2-sphere for $t \in (0, \varepsilon)$, that converges to g_0 on the *punctured* sphere as $t \downarrow 0$, and we can restrict this flow to the punctured sphere, which is conformally $\mathbb{R}^2$. The details and the claimed asymptotics can be found in [32, 34, 35].

In both Examples 3.1 and 3.2 we see incompleteness in the flow itself. In the theory we'll see later it turns out to be important to be able to flow *locally*, in which case even the initial data is incomplete. The following example shows how bad that can be.

Example 3.3 (Taken from [14]) For all $\varepsilon > 0$, there exists a Ricci flow $g(t)$ on the unit two-dimensional disc D in the plane, defined for $t \in [0, \varepsilon)$, such that $g(0)$ is the standard flat metric corresponding to the unit disc, but

$$\inf_D K_{g(t)} \to \infty \qquad \text{as } t \uparrow \varepsilon,$$

and $\mathrm{Vol}_{g(t)} D \to 0$ as $t \uparrow \varepsilon$.

If you've ever seen Perelman's famous Pseudolocality theorem before [25, §10], then you might mull over why the previous example does not provide a counterexample.

To construct Example 3.3, consider a 2-sphere that geometrically looks like a cylinder of length 2 and small circumference, with small hemispherical caps at the ends. See Fig. 3.1. We make the circumference small so that the total area is $8\pi\varepsilon \ll 1$. Thus the circumference of the cylinder is a little below $4\pi\varepsilon$. A theorem of Hamilton [17] tells us that the subsequent Ricci flow $\tilde{g}(t)$ will exist for a time ε, and at that time the volume converges to zero and the curvature blows up everywhere.

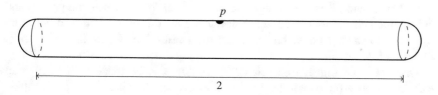

Fig. 3.1 Thin cylinder with caps

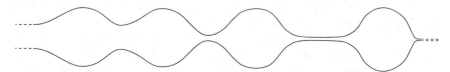

Fig. 3.2 $S^2 \times \mathbb{R}$ with thinner and thinner necks

Just before time ε, the surface looks like a tiny round sphere. Certainly it is easy to
see that the flow cannot exist any longer than this time since we can compute that

$$\frac{d}{dt}\,\mathrm{Vol}_{\tilde{g}(t)}\,S^2 = \int_{S^2}\frac{1}{2}\mathrm{tr}\frac{\partial\tilde{g}}{\partial t}dV_{\tilde{g}} = -2\int_{S^2}K_{\tilde{g}}\,dV_{\tilde{g}} = -8\pi,$$

by Gauss–Bonnet. To obtain the Ricci flow $g(t)$ on D of the example, we pull back
$\tilde{g}(t)$ under the exponential map $\exp_p$, defined with respect to the initial metric (not
the time t metric), restricted to the unit disc D in T_pM, where p is any point midway
along the cylinder.

Example 3.4 Conjecturally, we have an example of a smooth complete three-
manifold for which there does not exist a smooth Ricci flow solution even for a
short time. We construct this by connecting countably many three-spheres together
by necks that progressively become longer and thinner. Each neck should want
to pinch in a time that converges to zero as the necks get thinner and longer. See
Fig. 3.2 for a construction on $S^2 \times \mathbb{R}$. Closing up the left-hand side would give an
example on $\mathbb{R}^3$. This construction can be made even under the assumption that the
Ricci curvature is bounded below. Making this example rigorous is currently beyond
existing technology. One would need (for example) a pseudolocality theorem that is
valid in the presence of unbounded curvature, but this is unavailable currently.

3.3.2   What Is Known in the Noncompact Case

There are several situations in which we can say a lot about Ricci flows even though
the curvature can be unbounded.

The first theory to be developed was the theory for two-dimensional underlying
manifolds, which is also the one complex-dimensional Kähler case. This is the

easiest case, and gives the strongest results in terms of existence, uniqueness and asymptotic behaviour. The approach developed for this case has turned out to be important to understand the higher dimensional cases. We will see the main results in Sect. 3.3.3.

One way that higher dimensional Ricci flow can be made to have a clean existence theory for which there are no 'unnecessary' singularities, just as there are no unnecessary singularities in the general two-dimensional case, is to impose a sufficiently strong positivity of curvature condition. In particular, Cabezas-Rivas and Wilking [2] developed such a theory in the case that the initial manifold is complete with nonnegative complex sectional curvature.

Recently, several higher-dimensional cases that are substantially more singular have been treated. We will be looking particularly at the case in which the initial manifold is complete and three-dimensional, with a lower Ricci curvature bound. Note that Example 3.4 fits into this category, so we need to find a novel way of flowing. The solution is to flow locally, echoing the (quite different) work of Yang [38] and the concepts of Ricci flow on manifolds with boundary (see [12] and the earlier works referenced therein). Once a solution has been constructed, and we have strong-enough control on the evolution of the Ricci curvature and the distance function, we will obtain applications to the theory of Ricci limit spaces. See our work with Simon [19, 29] and Sect. 3.5 for the results, and Sect. 3.7 for some applications.

In [3], Bamler et al. describe a heat kernel method for controlling the curvature in certain cases, which allows the theory of [2] mentioned above to be generalised from nonnegative curvature to curvature bounded below, provided one assumes a global noncollapsing hypothesis. In [22], this work was combined with that of [29] to generalise both works. See also [24].

Finally, another interesting case in which to consider unbounded curvature is that of Kähler Ricci flow, though we will not attempt to survey that field here other than the one complex-dimensional case.

3.3.3   The Two-Dimensional Theory

In 2D we will be able to flow any smooth initial metric. We do not require the curvature to be bounded, so Shi's theorem 3.2 does not help. We do not even require that the initial metric is complete. Even if we can settle the existence problem, the level of generality here poses a potentially serious problem of nonuniqueness of solutions, and to understand this let us consider the PDE more carefully.

We already saw that in 2D, the Ricci flow is the equation

$$\frac{\partial g}{\partial t} = -2Kg,$$

where K is the Gauss curvature of g. If we choose local isothermal coordinates x, y, i.e. so the metric g can be written $e^{2u}(dx^2+dy^2)$ for a locally-defined scalar function

u, then we can write the Gauss curvature as $K = -e^{-2u} \Delta u$, where $\Delta = \frac{\partial^2}{\partial x^2} + \frac{\partial^2}{\partial y^2}$ is only locally defined. Thus the Ricci flow can be written locally as

$$\frac{\partial u}{\partial t} = e^{-2u} \Delta u, \tag{3.3}$$

which is the so-called logarithmic fast diffusion equation (up to a change of variables). Given such a parabolic equation, the accepted wisdom is that one should specify initial and boundary data in order to obtain existence and uniqueness. However here we abandon the idea of specifying boundary data in the traditional sense and instead replace it with the global condition that the flow should be complete for all positive times. Miraculously this completeness condition is just weak enough to permit existence, and just strong enough to force uniqueness.

Theorem 3.3 *Given any smooth (connected) Riemannian surface (M, g_0), possibly incomplete and possibly with unbounded curvature, define $T \in (0, \infty]$ depending on the conformal type of (M, g_0) by*

$$T := \begin{cases} \frac{1}{4\pi} \operatorname{Vol}_{g_0} M & \text{if } (M, g_0) \cong \mathbb{C} \text{ or } \mathbb{R}P^2, \\ \frac{1}{8\pi} \operatorname{Vol}_{g_0} M & \text{if } (M, g_0) \cong S^2, \\ \infty & \text{otherwise.} \end{cases}$$

Then there exists a Ricci flow $g(t)$ on M for $t \in [0, T)$ such that

1. *$g(0) = g_0$, and*
2. *$g(t)$ is complete for all $t \in (0, T)$.*

If $T < \infty$ then $\operatorname{Vol}_{g(t)} M \to 0$ as $t \uparrow T$. Moreover, this flow is unique in the sense that if $\tilde{g}(t)$ is any other complete Ricci flow on M with $\tilde{g}(0) = g_0$, existing for $t \in [0, \tilde{T})$, then $\tilde{T} \leq T$ and $\tilde{g}(t) = g(t)$ for all $t \in [0, \tilde{T})$.

The existence of this theorem was proved with Giesen in [13], following earlier work in [31]. The compact case, where the metric is automatically complete and the curvature bounded, was already proved by Hamilton and Chow [8, 17]. Other prior results include the extensive theory of the logarithmic fast diffusion equation, typically posed on $\mathbb{R}^2$, for example [10, 11, 36]. The uniqueness assertion was proved in [33] following a large number of prior partial results (see [33] for details).

We illustrate the theorem with two simple examples.

Example 3.5 Suppose (M, g_0) is the flat unit disc. Finding a Ricci flow starting with (M, g_0) is equivalent to solving (3.3) with the initial condition $u|_{t=0} = 0$. There is the obvious solution that has $u \equiv 0$ for all time, and we also saw the solution of Example 3.3. In general, standard PDE theory tells us that if we specify boundary data on the disc as a function of time (and space) then there exists a unique solution that approaches that data at the boundary. None of these solutions give complete metrics, so Theorem 3.3 is giving another solution that blows up at the boundary.

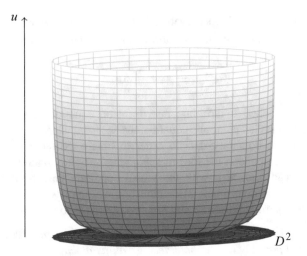

Fig. 3.3 Metric for positive time stretches at infinity in Example 3.5

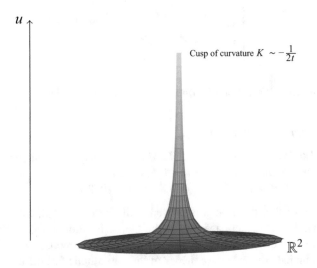

Fig. 3.4 Puncture turns into a hyperbolic cusp in Example 3.6

For this instantaneously complete flow, at small positive time $t > 0$ the flat metric gets adjusted asymptotically near the boundary to the Poincaré metric scaled to have constant curvature $-\frac{1}{2t}$, but remains almost flat on the interior. See Fig. 3.3.

Example 3.6 Alternatively, suppose that (M, g_0) is the flat Euclidean plane with one point removed to make it incomplete. This time the instantaneously complete solution stretches the metric immediately near the puncture to make it asymptotically like a hyperbolic cusp scaled to have curvature $-\frac{1}{2t}$. See Fig. 3.4.

It is important to appreciate that the flows themselves in Theorem 3.3 can have unbounded curvature at spatial infinity at each time [14] as well as initially. More precisely, the curvature can be unbounded above; it follows from [7] that our solutions always satisfy the Gauss curvature estimate $K_{g(t)} \geq -\frac{1}{2t}$. As alluded to in Remark 3.1, they might alternatively start as complete metrics of bounded curvature (so Shi's theorem 3.2 applies), develop blow-up of curvature in finite time (at spatial infinity) but then continue smoothly for all time [15]. These types of behaviour can even happen under strong positivity of curvature conditions, if one permits higher dimensional flows [2].

3.3.4   Proof Ideas for the 2D Theory

It will be helpful to outline some of the ideas in the proof of the existence of Theorem 3.3; we'll emphasise the ones that ended up being useful in the higher dimensional theory. A rough principle is that we would like to bring Shi's result into play despite the unboundedness of the curvature. We achieve this through the following steps, where we are free to assume that M is not closed (or we would be in the classical situation of Hamilton).

1. First we take an exhaustion of M by sets $\Omega_1 \subset \Omega_2 \subset \cdots \subset M$, each compactly contained in M and with smooth boundary.
2. Next, for each $i \in \mathbb{N}$ we restrict g_0 to Ω_i and blow up the metric conformally near the boundary to make the metric complete and of very large constant negative curvature near the boundary. Call the resulting metric g_i.
3. Since each (Ω_i, g_i) is complete, with bounded curvature, we can apply Shi's theorem 3.2 to obtain Ricci flows $g_i(t)$.
4. Next, we prove that each $g_i(t)$ exists for a uniform time (in fact for *all* time in this case) and enjoys uniform curvature estimates (independent of i).
5. The uniform estimates from the previous step give us compactness: a subsequence converges to a limit flow $g(t)$, this time on the whole of M.
6. Finally we argue that the limit flow $g(t)$ is complete for all positive times, and has g_0 as initial data.

We'll reuse the same strategy for higher dimensions, but the results are a little different.

3.4   Ricci Curvature and Ricci Limit Spaces

Most of the remaining material we will cover in these lectures will be geared to understanding Ricci flow in higher dimensions in the unbounded curvature case. The new ideas we wish to explain are best illustrated in the case of three-dimensional manifolds that have a lower Ricci curvature bound. These ideas generalise to higher dimensions if we assume correspondingly stronger curvature bounds. This theory has a number of potential applications to the understanding of the topology and geometry of manifolds with lower curvature bounds, but the application on which we focus in these lectures will be to so-called Ricci limit spaces, which we shall shortly define as some sort of limits of Riemannian manifolds with lower Ricci bounds. Before that, we survey some of the most basic consequences of a manifold having a lower Ricci bound.

3.4.1 Volume Comparison

Ricci curvature controls the growth of the volume of balls as the radius increases. The simplest result of this form is:

Theorem 3.4 (Bishop–Gromov, Special Case) *Given a complete Riemannian n-manifold (M, g) with $\mathrm{Ric} \geq 0$, and a point $x_0 \in M$, the* volume ratio *function*

$$r \mapsto \frac{\mathrm{VolB}_g(x_0, r)}{r^n} \tag{3.4}$$

is a weakly decreasing (i.e. nonincreasing) function of $r > 0$.

Since the limit of the function (3.4) as $r \downarrow 0$ is necessarily equal to ω_n, the volume of the unit ball in Euclidean n-space, this implies that $\mathrm{VolB}_g(x_0, r) \leq \omega_n r^n$. But of course, the theorem also gives us *lower* volume bounds: For $0 < r \leq R$, we have

$$\mathrm{VolB}_g(x_0, r) \geq (r/R)^n \, \mathrm{VolB}_g(x_0, R).$$

In fact, we do not even need to have the same centres here. Later it will be useful to observe that for any $x \in M$ we can set $R = r + 1 + d(x, x_0)$ and estimate, as in Fig. 3.5, that

$$\frac{\mathrm{VolB}_g(x, r)}{r^n} \geq R^{-n} \, \mathrm{VolB}_g(x, R) \geq R^{-n} \, \mathrm{VolB}_g(x_0, 1), \tag{3.5}$$

and thus knowing that one unit ball $\mathrm{VolB}_g(x_0, 1)$ has a specific lower volume bound then gives a lower bound for the volume of any other ball, though that lower bound gets weaker as x drifts far from x_0.

Theorem 3.4 also allows us to define a finite *asymptotic volume ratio*

$$\mathrm{AVR} := \lim_{r \to \infty} \frac{\mathrm{VolB}_g(x_0, r)}{r^n}.$$

In practice it is often particularly relevant whether AVR equals zero or is positive.

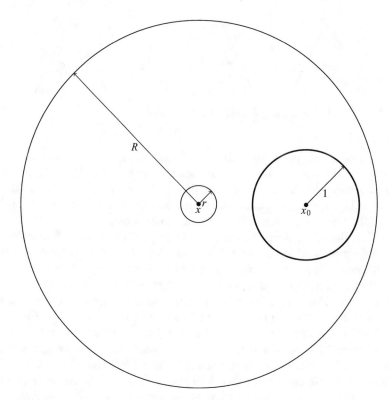

Fig. 3.5 Comparing the volumes of nonconcentric balls

Remark 3.2 The Bishop–Gromov theorem as stated extends easily to manifolds with more general Ricci lower bounds Ric $\geq -\alpha_0$, but now the volume of a ball of radius r can grow as fast as it does in the complete model space of constant sectional curvature where the constant is chosen so that the Ricci curvature equals $-\alpha_0$.

3.4.2   What Is a Ricci Limit Space?

Suppose that (M_i^n, g_i, x_i) is a sequence of pointed complete Riemannian manifolds, i.e. a sequence of complete Riemannian n-manifolds (M_i^n, g_i) together with points $x_i \in M_i$, and suppose that the sequence satisfies the properties that

$$\begin{cases} \mathrm{Ric}_{g_i} \geq -\alpha_0 \\ \mathrm{VolB}_{g_i}(x_i, 1) \geq v_0 > 0. \end{cases} \tag{3.6}$$

Fig. 3.6 Smoothed out cones approaching a Ricci limit space

Such a sequence of manifolds with lower Ricci curvature control has some sort of compactness property. That is, a subsequence will converge in a suitable sense. What it converges to will be more general than a Riemannian manifold, but will be a very special type of metric space. These limits will be what we mean by Ricci limit space. Before giving any details, let's imagine an example.

Consider the two-dimensional cone $C := (\mathbb{R}^2 \setminus \{0\}, dr^2 + \alpha r^2 d\theta^2)$, where $\alpha \in (0, 1)$ and we are working in polar coordinates. Intuitively there is a delta function of positive curvature at the vertex, i.e. at the origin, scaled by a factor $2\pi(1 - \alpha)$. If we view C as a metric space, and add in the origin, we have an example of a Ricci limit space with the origin as a singular point (these terms will be defined in a moment). This space will be the limit of a sequence of surfaces (M_i, g_i) that are constructed by smoothing out the cone point of C by a smaller and smaller amount as i gets large. Each of these surfaces can be assumed to have nonnegative Gauss curvature, and thus Ricci curvature bounded below. Moreover, if we let x_i be the origin for each i, we can arrange that $\text{VolB}_{g_i}(x_i, 1)$ converges to the area of the unit ball in C centred at the origin, i.e. to $\pi\alpha$, which is positive. Thus the conditions (3.6) will be satisfied, and we see that the cone C is a Ricci limit space; see Fig. 3.6.

So far this is very vague. The notion of convergence we are considering here is that of *pointed Gromov–Hausdorff* convergence. Before defining that, it makes sense to recall the usual notion of Gromov–Hausdorff convergence. A sequence of metric spaces (X_i, d_i) converges to a limit metric space (X, d) in the Gromov–Hausdorff sense if for all $\varepsilon > 0$ and for sufficiently large i (depending also on ε) there exists a map $f : X_i \to X$ such that

1. for all $x, y \in X_i$ we have

$$|d(f(x), f(y)) - d_i(x, y)| < \varepsilon,$$

2. for every $z \in X$, there exists $x \in X_i$ such that

$$d(f(x), z) < \varepsilon.$$

This is clearly a very weak notion of closeness. For example, a finer and finer lattice of points in the plane would converge to the plane itself.

We'll loosely call any such map f an ε-Gromov–Hausdorff approximation.

The general definition of *pointed* Gromov–Hausdorff convergence is similar: Given a complete metric space (X, d) and a point $x_\infty \in X$, i.e. a pointed metric space (X, d, x_∞), we say that a sequence of pointed metric spaces (X_i, d_i, x_i) converges to (X, d, x_∞) in the pointed Gromov–Hausdorff sense if given any radius $r > 0$ and any $\varepsilon > 0$, then for sufficiently large i (depending also on r and ε) there exists a map $f : B_{d_i}(x_i, r) \to X$ such that

1. $f(x_i) = x_\infty$,
2. for all $x, y \in B_{d_i}(x_i, r)$ we have

$$|d(f(x), f(y)) - d_i(x, y)| < \varepsilon,$$

3. the ε-neighbourhood of the image $f(B_{d_i}(x_i, r))$ contains the ball $B_d(x_\infty, r - \varepsilon) \subset X$.

In practice, we will essentially only consider this definition in the special case that the metric spaces (X_i, d_i, x_i) are in fact Riemannian manifolds (M_i^n, g_i, x_i). This restriction removes some annoying pathology by virtue of restricting us to working with length spaces and boundedly compact spaces (see [1] for definitions).

Indeed, working only with boundedly compact spaces, our limit (X, d, x_∞) is unique in the sense that any other (complete) limit will be isometric to this via an isometry that fixes the marked points [1, Theorem 8.1.7].

Another property that follows with this restriction to Riemannian manifolds, by virtue of restricting us to length spaces, is that for each $r > 0$, the balls $B_{g_i}(x_i, r)$ converge to the ball $B_d(x_\infty, r)$ in the Gromov–Hausdorff sense, which is not true in general. Beware that the converse fails: Gromov–Hausdorff convergence of open balls of arbitrary radius r does not imply the pointed convergence as given here. Nor does convergence of closed balls of arbitrary radius. (Just consider metric spaces each consisting of two points only.)

See [1, p. 271] ff for further details on this general topic.

Let's return to the original sequence of pointed manifolds satisfying conditions (3.6). By virtue of the lower Ricci curvature bound, and the resulting volume comparison discussed in Sect. 3.4.1, a compactness result of Gromov given in Theorem 3.12 of appendix tells us that we can pass to a subsequence in i and obtain pointed Gromov–Hausdorff convergence

$$(M_i, g_i, x_i) \to (X, d, x_\infty) \tag{3.7}$$

to some complete metric space (X, d) (even a length space) and $x_\infty \in X$.

Definition 3.1 A complete metric space (X, d) arising in a limit (3.7) of a sequence (M_i, g_i, x_i) of pointed complete Riemannian n-manifolds satisfying (3.6) is called a (noncollapsed) Ricci limit space.

In these lectures we will not consider so-called collapsed Ricci limit spaces, i.e. we will always assume the volume lower bound of (3.6).

Such Ricci limit spaces were studied extensively starting with Cheeger and Colding around the late 90s. They can be considered as rough spaces that can be viewed as having some sort of lower Ricci bound. As such they are analogues of Alexandrov spaces, which are metric spaces with a notion of lower *sectional* curvature bound. A synthetic notion of rough space with lower Ricci curvature bound (i.e. defined directly rather than as a limit of smooth spaces) is provided by the so-called RCD spaces, for example; see [37] for background. The exact link between Ricci limit spaces and RCD spaces is a topic of intense current study. One of the basic questions in that direction will be settled by the Ricci flow methods we will see in the next section; see Remark 3.4.

3.4.3   How Irregular Can a Ricci Limit Space Be?

Let's return to the example of a Ricci limit space that we have already seen, i.e. the two-dimensional cone C. Intuitively it has precisely one singular point; we now justify that by giving a definition of *singular point*. At each point p in a Ricci limit space (X, d), we can define a *tangent cone* to be a (pointed Gromov–Hausdorff) limit of rescalings $(X, \lambda_i d, p)$ for some $\lambda_i \to \infty$. We can always find a tangent cone by picking an arbitrary sequence $\lambda_i \to \infty$ and passing to a subsequence using Gromov compactness as discussed in Theorem 3.12. Strictly speaking we have only talked about Gromov compactness for sequences of Riemannian manifolds rather than metric spaces as here, but $(X, \lambda_i d, p)$ itself is the pointed Gromov–Hausdorff limit of $(M_j, \lambda_i^2 g_j, y_j)$ as $j \to \infty$, for some sequence of points $y_j \in M_j$, and the tangent cone can be viewed as a limit of rescaled and rebased pointed manifolds $(M_i, \mu_i g_i, y_i)$.

There may be different tangent cones corresponding to different sequences (see e.g. [5, Theorem 8.41]). Tangent cones have a lot of structure, see [4] for details.

Definition 3.2 A point p in a Ricci limit space (X, d) is singular if at least one tangent cone is not Euclidean space.

In fact, an elegant argument shows that if any one tangent cone is Euclidean space, then any other tangent cone (i.e. for any other sequence $\lambda_i \to \infty$) is Euclidean space too, see [4].

In our cone example, the tangent cones are Euclidean space away from the vertex (thus regular) and a nontrivial cone at the vertex (hence singular).

The singular set could be a lot worse in size, for example it could be dense. Let's see at least how one can construct a singular set with an accumulation point. To do this, start with a round sphere. Select a spherical cap that is strictly smaller than a hemisphere, and replace it with the top of a cone so that the surface is C^1, i.e. the tangent spaces match up. Thus we end up with an 'ice cream cone' as in Fig. 3.7.

We can now pick a countable collection of further spherical caps, of size converging to zero, that are mutually disjoint. Each of them can be replaced by a cone as before. As the caps become smaller, the cones become 'flatter'. Each cone vertex will be a singular point. Clearly we have constructed another Ricci limit space, this time with accumulating singular points, which we call (X_0, d_0).

Fig. 3.7 Ice cream cone

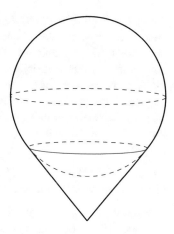

Although the singular set here is infinite, it nevertheless has Hausdorff dimension zero, which is optimal:

Theorem 3.5 (Cheeger and Colding [5]) *A noncollapsed Ricci limit space has Hausdorff dimension n and its singular set has Hausdorff dimension no larger than n − 2.*

By considering the product of (X_0, d_0) with $\mathbb{R}^{n-2}$, which is clearly another Ricci limit space, we see that the control on the size of the singular set of Theorem 3.5 is optimal in all dimensions.

Our example (X_0, d_0) may have an infinite singular set, but it is *topologically regular* in the sense that it is *homeomorphic* to a smooth manifold, in this case S^2. The simple conical points present in this example have neighbourhoods homeomorphic to flat discs. However even this milder form of regularity fails if we increase the dimension from 2 to 4 because of the following example.

Example 3.7 There exists a four-dimensional complete Riemannian manifold (M, g_0), known as the Eguchi–Hanson manifold, that is Ricci-flat, and which looks asymptotically like $\mathbb{R}^4/\sim$, where $\sim$ is the equivalence relation that identifies each point $x \in \mathbb{R}^4$ with $-x$. By picking an arbitrary point $x_0 \in M$ and taking pointed manifolds $(M_i, g_i, x_i) := (M, \frac{1}{i}g_0, x_0)$ for each i, which satisfy the conditions (3.6), and converge in the pointed Gromov–Hausdorff sense to $\mathbb{R}^4/\sim$, we find that $\mathbb{R}^4/\sim$ is a Ricci limit space. But this is a cone over $\mathbb{R}P^3$, and being a cone over something other than a sphere, it cannot be homeomorphic to a manifold.

Remark 3.3 The term *cone* can refer to many different things. Here when we say cone over $\mathbb{R}P^3$ we can start with the Riemannian cone defined on $(0, \infty) \times \mathbb{R}P^3$, with the warped product metric $dr^2 + r^2 g_1$, where $r \in (0, \infty)$ and g_1 is the standard metric on $\mathbb{R}P^3$, then view the space as a metric space and complete it by adding one point at the tip of the cone.

By taking a product of the Eguchi–Hanson manifold and $\mathbb{R}^{n-4}$, we see that in general the best regularity one can hope for is for our Ricci limit space to be homeomorphic to a manifold off a set of dimension $n - 4$. It is a long-standing conjecture, normally attributed to Anderson, Cheeger, Colding and Tian (ACCT) that in some sense this is the case. We'll give a precise formulation of the conjecture in general dimension in a moment, once we have introduced some more theory. But we can immediately state it precisely in the three-dimensional case: The whole Ricci limit space should be homeomorphic to a topological manifold. With Miles Simon we prove more:

Theorem 3.6 (With Simon [29]) *In the case $n = 3$, any Ricci limit space is locally bi-Hölder homeomorphic to a smooth manifold.*

The proof of this will require a string of new ideas, some of which are due to Hochard [19]. In fact, a slight improvement is possible.

Theorem 3.7 (With McLeod [23], Using Mainly the Theory from [28, 29]) *In the case $n = 3$, any Ricci limit space is globally homeomorphic to a smooth manifold via a homeomorphism that is locally bi-Hölder.*

Prior to these results, it was only possible to give topological regularity statements about the *regular set* of a Ricci limit space (i.e. the complement of the singular set) or suitable generalisations thereof. This reduces considerations to spaces that are almost-Euclidean in some sense, which is a completely different task. However, it does have the advantage of having a chance of working in general dimension.

Theorem 3.8 (Cheeger and Colding [5]) *There exists an open neighbourhood of the regular set of a general Ricci limit space that is locally bi-Hölder homeomorphic to a smooth manifold.*

In Theorem 3.8 we can make the Hölder exponent as close as we like to 1 by shrinking the set (still containing the regular set). It is unknown whether we can make the homeomorphism Lipschitz.

We now return to formulate the ACCT conjecture in general dimension. For each $k \in \{0, \ldots, n - 1\}$, consider the subset S_k of the singular set S consisting of points at which no tangent cone can be written as a product of $\mathbb{R}^{k+1}$ with something else. (At most they can be a product of $\mathbb{R}^k$ with something else.) Note that

$$S_0 \subset S_1 \subset \cdots \subset S_{n-1} = S.$$

Cheeger–Colding's theory [5] tells us that the Hausdorff dimension of S_k is at most k. As usual, we are considering noncollapsed Ricci limit spaces [5]. In fact, in this case, Cheeger–Colding also proved that $S_{n-2} = S$, which is how one proves Theorem 3.5.

The full ACCT conjecture is:

Conjecture 3.1 ([5, Conjecture 0.7].) Given any Ricci limit space X, the interior of $X \setminus S_{n-4}$ is homeomorphic to a topological manifold.

Remark 3.4 One consequence of Theorem 3.6 is that we can settle the most obvious questions concerning whether synthetically-defined metric spaces with lower Ricci bounds always arise as noncollapsed Ricci limit spaces. One can see the cone over $\mathbb{R}P^2$ (plus the vertex) as such a metric space with non-negative Ricci curvature (e.g. a $RCD^*(0, 3)$ space, as shown in [20]) but it is not a manifold, being a cone over something other than a sphere, and thus cannot be a Ricci limit space by our theorem. More significantly, our theory in [29] gives a local distinction in that a unit ball centred at the vertex of the cone cannot arise as the pointed Gromov–Hausdorff limit of a sequence of possibly incomplete Riemannian manifolds (M_i^3, g_i, x_i) each with the property that $B_{g_i}(x_i, 1) \subset\subset M_i$ (i.e. each manifold can fit in a unit ball) and $\mathrm{Ric}_{g_i} \geq 0$, say. Note that it is much easier (and much weaker) to show that the cone over $\mathbb{R}P^2$ cannot arise as the pointed Gromov–Hausdorff limit of a sequence of *closed* Riemannian 3-manifolds each with $\mathrm{Ric}_{g_i} \geq 0$. A global restriction of this form imposes very strong rigidity on the geometry and topology of the approximating closed manifolds. It is not even clear that a *smooth* complete Riemannian 3-manifold with non-negative Ricci curvature can be approximated by closed 3-manifolds with the same curvature restriction.

3.4.4 How to Apply Ricci Flow to Ricci Limit Spaces

The central question when trying to apply Ricci flow to Ricci limit spaces is:

> Can we start the Ricci flow with a Ricci limit space as initial data?

Clearly we will have to revisit the traditional viewpoint of how to pose Ricci flow that we described at the start of the lectures. Now, when we are given initial data that is a metric space (X, d) rather than a Riemannian manifold, we are asking Ricci flow to create not only a one-parameter family of metrics $g(t)$, but also the underlying manifold on which they live! The family $g(t)$ will now only live for positive times (i.e. not at $t = 0$) and we will want $(M, d_{g(t)})$ to converge in a sufficiently strong sense to the initial data (X, d) as $t \downarrow 0$.

Slightly more precisely, we will want that the distance metric $d_{g(t)}$ converges nicely to a distance metric d_0 on M as $t \downarrow 0$, and that (M, d_0) is isometric to the Ricci limit space (X, d). Here *nice* convergence will be more than local uniform convergence on the domain $M \times M$ of the distance metrics.

Luckily we are not trying to flow a general metric space, but instead a Ricci limit space, which has a lot more structure. We can therefore see a potential strategy that essentially follows the strategy that worked in the 2D case. (We will need more ideas

from the 2D case in due course.) For a Ricci limit space (X, d) arising as the pointed
limit of (M_i, g_i, x_i) we might try to:

1. Flow each (M_i, g_i) to give Ricci flows $(M_i, g_i(t))$, and argue that they exist on
   a time interval $[0, T]$ that is independent of i. (T is allowed to depend on α_0 and
   v_0.)
2. Derive uniform estimates on the full curvature tensor of $g_i(t)$ that are independent of i, but can depend on $t \in (0, T]$ as well as α_0 and v_0.
3. Use parabolic regularity to improve these bounds to bounds on all derivatives of
   the full curvature tensor.
4. Appeal to the smooth 'Cheeger–Gromov–Hamilton' compactness that one
   obtains from these curvature bounds. More precisely, after passing to a
   subsequence, there exist a smooth manifold M, a smooth Ricci flow $g(t)$ for
   $t \in (0, T]$ and a point $x_\infty \in M$ such that

$$(M_i, g_i(t), x_i) \to (M, g(t), x_\infty)$$

   smoothly as $i \to \infty$. Smooth convergence means that there exist an exhaustion
   $x_\infty \in \Omega_1 \subset \Omega_2 \subset \cdots \subset M$ and diffeomorphisms $\varphi_i : \Omega_i \to M_i$ such that
   $\varphi_i(x_\infty) = x_i$ and

$$\varphi_i^* g_i(t) \to g(t)$$

   smoothly locally on $M \times (0, T]$. See [30, §7] for a more detailed description of
   Cheeger–Gromov convergence.
5. Show that the distance metric $d_{g(t)}$ converges to some limit metric d_0 at least
   uniformly as $t \downarrow 0$. (In practice, stronger convergence will be required.)
6. Show that having passed to a subsequence, the sequence (M_i, g_i, x_i) converges
   in the pointed Gromov–Hausdorff sense to (M, d_0, x_∞). But this sequence also
   converges to the Ricci limit space we are trying to flow, so that must be isometric
   to (M, d_0, x_∞).

Note that the smooth manifold M that we require has been *created* by the Ricci flow.

If this programme could be completed, then we would have succeeded in starting
the Ricci flow with a Ricci limit space. Additionally, if we could prove that the
distance metric $d_{g(t)}$ is sufficiently close to d_0, depending on t, in a strong enough
sense, we can hope that the identity map from (M, d_0) to $(M, d_{g(t)})$, i.e. to the
smooth Riemannian manifold $(M, g(t))$, will be bi-Hölder. This would give a proof
of Theorem 3.6.

Unfortunately, this strategy is a little naive as written. Even the first step fails
at the first hurdle: Example 3.4 showed that we cannot expect to even start the
Ricci flow with a general smooth complete manifold even if the Ricci curvature
is bounded below. Miles Simon carried the programme through in the case that
the approximating manifolds are closed [27] so this problem doesn't arise. For the
general case this problem forces us to work locally. Once we do that, the strategy
will work.

3.5   Local Ricci Flow in 3D

3.5.1   Local Ricci Flow Theorem with Estimates

The following theorem, proved with Miles Simon, gives the local Ricci flow that we need. We can apply it to each of the (M_i, g_i, x_i) considered in the previous section in order to make the strategy there work. As mentioned earlier, local Ricci flow under different hypotheses or with weaker conclusions on curvature and distance has been considered by several authors including D. Yang [38], P. Gianniotis [12], and R. Hochard [19].

Theorem 3.9 (Special Case of Results in [29]) *Suppose (M^3, g_0) is a complete Riemannian manifold, and $x_0 \in M$ so that*

$$
\begin{cases}
\mathrm{Ric}_{g_0} \geq -\alpha_0 \\
\mathrm{VolB}_{g_0}(x_0, 1) \geq v_0 > 0.
\end{cases}
\tag{3.8}
$$

Then for all $r \geq 2$ there exists a Ricci flow $g(t)$ (incomplete) defined on $B_{g_0}(x_0, r)$ for $t \in [0, T]$, with $g(0)$ equal to (the restriction of) g_0 such that

$$
B_{g(t)}(x_0, r-1) \subset\subset B_{g_0}(x_0, r),
\tag{3.9}
$$

and with Ricci and volume bounds persisting in the sense that

$$
\begin{cases}
\mathrm{Ric}_{g(t)} \geq -\alpha \\
\mathrm{VolB}_{g(t)}(x_0, 1) \geq v > 0,
\end{cases}
\tag{3.10}
$$

but additionally so that

$$
|\mathrm{Rm}|_{g(t)} \leq \frac{c_0}{t},
\tag{3.11}
$$

where $T, \alpha, v, c_0 > 0$ depend only on α_0, v_0 and r. Moreover, we have distance function convergence $d_{g(t)} \to d_{g_0}$ as $t \downarrow 0$ in the sense that (for example) for all $x, y \in B_{g_0}(x_0, r/8)$ we have

$$
d_{g_0}(x, y) - \beta \sqrt{c_0 t} \leq d_{g(t)}(x, y) \leq e^{\alpha t} d_{g_0}(x, y)
\tag{3.12}
$$

and

$$
d_{g_0}(x, y) \leq \gamma [d_{g(t)}(x, y)]^{\frac{1}{1+4c_0}},
\tag{3.13}
$$

where β is universal, α is as above, and γ depends only on c_0, thus on α_0, v_0 and r.

Example 3.3 illustrates the importance of (3.9) in the context of incomplete Ricci flows. Our flows have stable balls that do not instantly collapse. This property, together with the refined information from (3.12) and (3.13) relies crucially on the curvature estimates of (3.10) and (3.11). In fact, it will be useful to split off what control one obtains, in slightly greater generality, in the following remark.

Remark 3.5 (Distance Estimates) Given a Ricci flow $g(t)$ of arbitrary dimension n, defined only for *positive* times $t \in (0, T]$, and satisfying the curvature bounds

$$
\begin{cases}
\mathrm{Ric}_{g(t)} \geq -\alpha \\
\mathrm{Ric}_{g(t)} \leq \dfrac{(n-1)c_0}{t}
\end{cases}
\tag{3.14}
$$

we automatically obtain curvature estimates like in Theorem 3.9 that are strong enough to extend the distance metric to $t = 0$. Indeed, for any $0 < t_1 \leq t_2 \leq T$, we have

$$
d_{g(t_1)}(x, y) - \beta\sqrt{c_0}(\sqrt{t_2} - \sqrt{t_1}) \leq d_{g(t_2)}(x, y) \leq e^{\alpha(t_2 - t_1)} d_{g(t_1)}(x, y),
\tag{3.15}
$$

where $\beta = \beta(n)$. In particular, the distance metrics $d_{g(t)}$ converge locally uniformly to some distance metric d_0 as $t \downarrow 0$, and

$$
d_0(x, y) - \beta\sqrt{c_0 t} \leq d_{g(t)}(x, y) \leq e^{\alpha t} d_0(x, y),
\tag{3.16}
$$

for all $t \in (0, T]$. Furthermore, there exists $\eta > 0$ depending on n, c_0 and the size of the region in space-time we are considering such that

$$
d_{g(t)}(x, y) \geq \eta \left[d_0(x, y) \right]^{1 + 2(n-1)c_0},
\tag{3.17}
$$

for all $t \in (0, T]$. Here we are being vague as to where the points x, y live, and the dependencies of η. In the local version of this result that we need, these issues require some care. See [29, Lemma 3.1] for full details.

Remark 3.6 The distance estimates of Theorem 3.9 imply that over appropriate local regions Ω, we have $id : (\Omega, d_{g(t)}) \to (\Omega, d_{g_0})$ is Hölder and $id : (\Omega, d_{g_0}) \to (\Omega, d_{g(t)})$ is Lipschitz. Similar statements follow from the estimates of Remark 3.5, with d_0 in place of d_{g_0}. In particular, the identity is a bi-Hölder homeomorphism from the potentially-rough initial metric to the positive-time smooth metric.

Remark 3.7 (Parabolic Rescaling of Ricci Flows) It will be essential to digest the natural parabolic rescaling of a Ricci flow to obtain a new Ricci flow. Given a Ricci flow $g(t)$ and a constant $\lambda > 0$, we can define a scaled Ricci flow

$$
g_\lambda(t) := \lambda g(t/\lambda),
$$

on the appropriately scaled time interval. When we do this rescaling, the curvature is scaled, for example the sectional curvatures are all multiplied by a factor λ^{-1}. One significant aspect of the uniform curvature estimate (3.11) obtained in Theorem 3.9 is that it is *invariant* under rescaling. That is, we get the same estimate for the rescaled flow *with the same* c_0.

Remark 3.8 A well-known consequence of the curvature decay (3.11), via parabolic regularity, known as Shi's estimates in this context [18, §13], is that for all $k \in \mathbb{N}$ we have

$$|\nabla^k \mathrm{Rm}|_{g(t)} \leq \frac{C}{t^{1+k/2}}, \tag{3.18}$$

where C is allowed to depend on k as well as α_0, v_0 and r.

3.5.2   Getting the Flow Going

The target of Theorem 3.9 is to start the Ricci flow locally for a controlled amount of time, with a selection of quantitative estimates. But first we want to focus on the problem of merely starting the flow locally, without estimates and for an uncontrolled time interval. We follow the general strategy we used in the 2D theory, as in Sect. 3.3.3. To flow on a ball $B_{g_0}(x_0, r)$ as in the theorem, we would like to try to conformally blow up the metric in order to obtain a complete metric with constant negative curvature asymptotically, and hence one of bounded curvature. In 2D, this construction is straightforward [31]. In higher dimensions we use a lemma of Hochard [19], developed precisely for this purpose, following Simon [27, Theorem 8.4].

The following lemma basically says that given a local region of a Riemannian manifold with curvature bounded by $|\mathrm{Rm}| \leq \rho^{-2}$ on that local region, we can conformally blow up the metric near the boundary so that the resulting metric is *complete* and has curvature bounded by $|\mathrm{Rm}| \leq \gamma \rho^{-2}$, where the factor γ only depends on the dimension. It is asymptotically like a scaled hyperbolic metric.

Lemma 3.2 (Variant of Hochard, [19, Lemma 6.2]) *Let (N^n, g) be a smooth Riemannian manifold (that need not be complete) and let $U \subset N$ be an open set. Suppose that for some $\rho \in (0, 1]$, we have $\sup_U |\mathrm{Rm}|_g \leq \rho^{-2}$, $B_g(x, \rho) \subset\subset N$ and $\mathrm{inj}_g(x) \geq \rho$ for all $x \in U$. Then there exist an open set $\tilde{U} \subset U$, a smooth metric $\tilde{g}$ defined on $\tilde{U}$ and a constant $\gamma = \gamma(n) \geq 1$, such that each connected component of $(\tilde{U}, \tilde{g})$ is a complete Riemannian manifold such that*

(1) $\sup_{\tilde{U}} |\mathrm{Rm}|_{\tilde{g}} \leq \gamma \rho^{-2}$,
(2) $U_\rho \subset \tilde{U} \subset U$, and
(3) $\tilde{g} = g$ on $\tilde{U}_\rho \supset U_{2\rho}$,

where $U_s = \{x \in U \mid B_g(x, s) \subset\subset U\}$.

In order to achieve our modest task of at least starting the flow locally on a ball $B_{g_0}(x_0, r)$, we can appeal to Lemma 3.2 with (N, g) of the lemma equal to (M, g_0), with U of the lemma equal to $B_{g_0}(x_0, r + 2) \subset M$, and with $\rho \in (0, 1]$ sufficiently small so that $\sup_U |\mathrm{Rm}|_g \leq \rho^{-2}$ and $\mathrm{inj}_g(x) \geq \rho$ for all $x \in U$. The output of the lemma is a complete Riemannian manifold $(\tilde{U}, \tilde{g})$ of bounded curvature, where $\tilde{U} \subset M$, such that $\tilde{g} = g_0$ on $B_{g_0}(x_0, r)$. To obtain the local Ricci flow, we simply apply Shi's theorem 3.2 to $(\tilde{U}, \tilde{g})$ to obtain a Ricci flow $g(t)$, $t \in [0, T]$, and then restrict the flow to $B_{g_0}(x_0, r)$.

We have succeeded in starting the flow, but there is a major problem that the existence time T depends on $\sup_{\tilde{U}} |\mathrm{Rm}|_{\tilde{g}}$, which in turn depends on $\sup_{B_{g_0}(x_0, r+2)} |\mathrm{Rm}|_{g_0}$. This is not permitted. To be useful, the existence time can only ultimately depend on a lower Ricci bound, a lower bound on the volume of $B_{g_0}(x_0, 1)$ and the radius r.

To resolve this problem, we need to extend the flow we have found. This will be achieved with the extension lemma of the following section, after which we'll pick up the proof of Theorem 3.9 in Sect. 3.5.4.

3.5.3   The Extension Lemma

The extension lemma will tell us that if we have a reasonably controlled Ricci flow on some local region $B_{g_0}(x_0, r_1)$, with reasonably controlled initial metric, then we can extend the Ricci flow to a longer time interval, albeit on a smaller ball, whilst retaining essentially the same estimates on the extension. This will be iterated to prove Theorem 3.9.

Lemma 3.3 (Extension Lemma [29]) *Given $v_0 > 0$, there exist $c_0 \geq 1$ and $\tau > 0$ such that the following is true. Let $r_1 \geq 2$, and let (M, g_0) be a smooth three-dimensional Riemannian manifold with $B_{g_0}(x_0, r_1) \subset\subset M$, and*

(i) $\mathrm{Ric}_{g_0} \geq -\alpha_0$ for some $\alpha_0 \geq 1$ on $B_{g_0}(x_0, r_1)$, and
(ii) $\mathrm{Vol} B_{g_0}(x, s) \geq v_0 s^3$ for all $s \leq 1$ and all $x \in B_{g_0}(x_0, r_1 - s)$.

Assume additionally that we are given a smooth Ricci flow $(B_{g_0}(x_0, r_1), g(t))$, $t \in [0, \ell_1]$, where $\ell_1 \leq \frac{\tau}{200\alpha_0 c_0}$, with $g(0)$ equal to the restriction of g_0, for which

(a) $|\mathrm{Rm}|_{g(t)} \leq \frac{c_0}{t}$ and
(b) $\mathrm{Ric}_{g(t)} \geq -\frac{\tau}{\ell_1}$

on $B_{g_0}(x_0, r_1)$ for all $t \in (0, \ell_1]$. Then, setting $\ell_2 = \ell_1(1 + \frac{1}{4c_0})$ and $r_2 = r_1 - 6\sqrt{\frac{\ell_2}{\tau}} \geq 1$, the Ricci flow $g(t)$ can be extended smoothly to a Ricci flow over the longer time interval $t \in [0, \ell_2]$, albeit on the smaller ball $B_{g_0}(x_0, r_2)$, with

(a') $|\mathrm{Rm}|_{g(t)} \leq \frac{c_0}{t}$ and
(b') $\mathrm{Ric}_{g(t)} \geq -\frac{\tau}{\ell_2}$

throughout $B_{g_0}(x_0, r_2)$ for all $t \in (0, \ell_2]$.

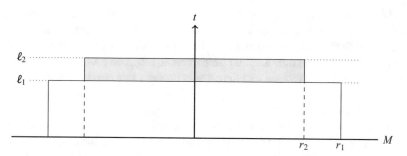

Fig. 3.8 Domain of original Ricci flow In the Extension Lemma 3.3 and (shaded) additional domain to which we extend

See Fig. 3.8 for the domains where the Ricci flows are defined.

3.5.4   Using the Extension Lemma to Construct a Local Ricci Flow

Assuming the Extension lemma, we can now pick up from where Sect. 3.5.2 ended, and give a sketch of the proof of the existence and curvature bounds of Theorem 3.9. See the end of Section 4 of [29] for the fine details. For the estimates on the distance function, the crucial thing is to have the lower Ricci and upper Rm bounds of the theorem. The estimates then follow from [29, Lemma 3.1], cf. Remark 3.5.

Before we start, it will be useful to reformulate the conditions (3.8) of Theorem 3.9 a little. The claim is that taking r from Theorem 3.9, the conditions (3.8) imply that for $r_1 = r + 1$,

$$\mathrm{VolB}_{g_0}(x, s) \geq \tilde{v}_0 s^3 \quad \text{for all } s \leq 1 \text{ and all } x \in B_{g_0}(x_0, r_1 - s), \tag{3.19}$$

where $\tilde{v}_0 > 0$ depends on v_0, α_0 and r. To prove this is a simple comparison geometry argument; if we had (3.8) with $\alpha_0 = 0$, then this was done in Sect. 3.4.1— see (3.5) in particular. The general case is a minor variant.

This observation means that we may as well replace the second condition of (3.8) by the condition (3.19). We will do that below, dropping the tilde on $\tilde{v}_0$. Of course, this means that we satisfy the conditions (i) and (ii) of Lemma 3.3.

Now we have this new v_0, we can use it to obtain constants c_0 and τ from Extension lemma 3.3.

The next step is to get the flow going as in Sect. 3.5.2, except that we construct the flow on the ball $B_{g_0}(x_0, r + 1)$ of slightly larger radius than before. This flow will exist on some nontrivial but otherwise uncontrolled time interval $[0, \ell_1]$. By reducing $\ell_1 > 0$ if necessary we can ensure that conditions (a) and (b) of the Extension lemma are satisfied.

At this point, if $\ell_1 > \frac{\tau}{200\alpha_0 c_0}$, then we have existence of a Ricci flow on a time interval with the right dependencies. But in general, we will have existence for an uncontrollably short time interval. In this latter case, we apply the Extension lemma. This gives us existence over a longer time interval, albeit on a smaller ball, and with the same estimates. Thus we can iterate this process to extend the flow more and more.

We stop iterating only if either ℓ_1 is large enough as above, or if the radius of the ball on which we get estimates becomes smaller than the radius r on which we would like our ultimate solution. Here is a potential worry: If the original existence time were extremely small, then we would need a huge number of iterations to get a reasonable existence time. Each time we extend on a smaller ball, so there is a concern that the whole space is quickly exhausted as the balls get too small. Luckily, the smaller the original existence time, the smaller the reduction in the size of the ball need be. It turns out that we can ask that the radius of the balls where we obtain existence never drops below r, but still be guaranteed a good time of existence.

Section 4 of [29] gives the details.

3.5.5 Proof of the Extension Lemma 3.3

We sketch some of the main ingredients and ideas of the proof of the Extension lemma. The objective is to make the full proof in [29, §4] accessible.

The first result expresses that a Ricci flow that has an initial volume lower bound, and a Ricci lower bound for each time, must regularise itself over short time periods. The lemma has many variants; in particular, a global version can be found in the earlier work of Simon [27].

Lemma 3.4 (Local Lemma, cf. [28, Lemma 2.1]) *Let* $(N^3, g(t))$, *for* $t \in [0, T]$, *be a Ricci flow such that for some fixed* $x \in N$ *we have* $B_{g(t)}(x, 1) \subset\subset N$ *for all* $t \in [0, T]$, *and so that*

(i) $\mathrm{Vol}B_{g(0)}(x, 1) \geq v_0 > 0$, *and*
(ii) $\mathrm{Ric}_{g(t)} \geq -1$ *on* $B_{g(t)}(x, 1)$ *for all* $t \in [0, T]$.

Then there exist $C_0 \geq 1$ *and* $\hat{T} > 0$, *depending only on* v_0, *such that* $|\mathrm{Rm}|_{g(t)}(x) \leq C_0/t$, *and* $\mathrm{inj}_{g(t)}(x) \geq \sqrt{t/C_0}$ *for all* $0 < t \leq \min(\hat{T}, T)$.

The key point to digest here is that we obtain the curvature decay $|\mathrm{Rm}|_{g(t)} \leq C_0/t$, which is an estimate that is invariant under parabolic rescaling of the flow as in Remark 3.7. The main idea of the proof is as follows. If the curvature estimate failed in a shorter and shorter time and with disproportionately larger and larger C_0, for some sequence of flows, then we can parabolically rescale each flow so that they exist for a longer and longer time, now with almost nonnegative Ricci curvature, but still end up with $|\mathrm{Rm}| = 1$ at the end. It then makes sense to pass to a limit of the flows (in the smooth Cheeger–Gromov sense, see, for example, [30, §7] and the discussion in Sect. 3.4.4) and obtain a limit smooth Ricci flow that existed since time $-\infty$, has Ric ≥ 0, and has $|Rm| = 1$ at time 0, say, but which has bounded curvature. Moreover it is possible to argue that the hypothesis $\mathrm{Vol}B_{g(0)}(x, 1) \geq v_0 > 0$ eventually tells us that this limit Ricci flow has positive asymptotic volume ratio in the sense that the volume of a ball of radius $r > 0$ is at least εr^3 for some $\varepsilon > 0$, however large we take r. Finally, a result of Chow and Knopf [9, Corollary 9.8] tells us that such a 3D complete ancient Ricci flow with bounded curvature must have nonnegative sectional curvature.

This puts our limit flow in the framework of Perelman's famous κ-solutions [25, §11] that are pivotal in understanding the structure of Ricci flow singularities. The output of Perelman's work is that the asymptotic volume ratio of such a nonflat solution must in fact be *zero*, contradicting what we already saw, and completing the (sketch) proof. See [28, §5] for the argument in full detail.

The following key ingredient gives a local version of the preservation of nonnegative Ricci curvature that we saw at the beginning of these lectures. The Ricci lower bounds it provides will be critical in the applications we have in mind.

Lemma 3.5 (cf. Double Bootstrap Lemma [28, Lemma 9.1]) *Let $(N^3, g(t))$, for $t \in [0, T]$, be a Ricci flow such that $B_{g(0)}(x, 2) \subset\subset N$ for some $x \in N$, and so that throughout $B_{g(0)}(x, 2)$ we have*

(i) $|Rm|_{g(t)} \leq \frac{c_0}{t}$ for some $c_0 \geq 1$ and all $t \in (0, T]$, and
(ii) $\mathrm{Ric}_{g(0)} \geq -\delta_0$ for some $\delta_0 > 0$.

Then there exists $\hat{S} = \hat{S}(c_0, \delta_0) > 0$ such that for all $0 \leq t \leq \min(\hat{S}, T)$ we have

$$\mathrm{Ric}_{g(t)}(x) \geq -100\delta_0 c_0.$$

Equipped with these two lemmata, we will attempt to outline some of the central ideas of the Extension lemma 3.3. Amongst the various details we suppress is the estimation of the exact domain on which we can extend the flow. For the precise details of the proof, see Section 4 of [29].

Step 1: Choose the Constants
Let v_0 be the constant from the Extension lemma 3.3, and plug it in to the Local lemma 3.4 to obtain constants C_0 and $\hat{T}$. Extracting the constant $\gamma \geq 1$ from Hochard's lemma 3.2, we are already in a position to specify $c_0 := 4\gamma C_0$.

In turn this allows us to specify $\delta_0 := \frac{1}{100c_0}$, and appeal to the Double Bootstrap lemma 3.5 to obtain $\hat{S}$. We can then insist to work on time intervals $[0, \tau]$ where $\tau \leq \hat{T}, \hat{S}$ so that all our supporting lemmata apply.

Step 2: Improvement of the Curvature Decay
The flow we are given in the Extension lemma already enjoys c_0/t curvature decay, where we recall that $c_0 := 4\gamma C_0 > C_0$. By parabolically scaling up our flow, scaling up time by a factor τ/ℓ_1, and distances by a factor $\sqrt{\tau/\ell_1}$, our flow ends up with a lower Ricci bound of -1, and the Local lemma 3.4 applies over all unit balls to give C_0/t curvature bounds (and injectivity radius bounds) for $t \in (0, \tau]$. Scaling back, we get again C_0/t curvature bounds for the original flow, holding now up until time $\tau \times \frac{\ell_1}{\tau} = \ell_1$, as desired.

Hidden a little here is that the details of this step require the assumed c_0/t curvature decay in order to verify the hypotheses of the Local lemma.

Step 3: Continuation of the Flow

The previous step has obtained improved control on the curvature of the originally given Ricci flow on an appropriate interior region of $B_{g_0}(x_0, r_1)$, and in particular at time ℓ_1 we have

$$|\text{Rm}|_{g(\ell_1)} \leq \frac{C_0}{\ell_1}.$$

We can now follow Hochard, Lemma 3.2, and modify the metric around the outside in order to make it complete and with a curvature bound

$$|\text{Rm}| \leq \gamma \frac{C_0}{\ell_1} = \frac{c_0}{4\ell_1},$$

by definition of c_0. Having a complete, bounded curvature metric, we can apply Shi's theorem 3.2 to extend the flow for a time $\frac{1}{16} \times \frac{4\ell_1}{c_0} = \frac{\ell_1}{4c_0}$, maintaining a curvature bound

$$|\text{Rm}| \leq 2 \times \frac{c_0}{4\ell_1} \leq \frac{c_0}{\ell_2},$$

where $\ell_2 := \ell_1(1 + \frac{1}{4c_0}) \leq 2\ell_1$. Thus we have proved a little more than

$$|\text{Rm}|_{g(t)} \leq \frac{c_0}{t}$$

for the extended Ricci flow $g(t)$, which is part (a') of the desired conclusion.

Step 4: Ricci Lower Bounds

Now we have indirectly obtained c_0/t curvature decay for the extended flow, we are in a position to obtain the crucial lower Ricci bound conclusion of part (b'). Similarly to before, we parabolically scale up the extended flow, scaling time by a factor τ/ℓ_2 so that the new flow lives for a time τ. This preserves our c_0/t curvature decay, but improves the initial Ricci lower bound from $-\alpha_0$ to

$$-\alpha_0 \frac{\ell_2}{\tau} \geq -2\alpha_0 \frac{\ell_1}{\tau} \geq -\frac{1}{100c_0} = -\delta_0.$$

Thus we can apply the Double Bootstrap lemma 3.5 to conclude that the rescaled flow has a Ricci lower bound of $-100c_0\delta_0 = -1$. Rescaling back, we obtain that our original extended flow has a Ricci lower bound of $-\tau/\ell_2$ as required.

This completes the sketch of the proof of the Extension lemma 3.3.

Remark 3.9 The Pyramid Ricci flow construction that we briefly discuss in Sect. 3.6 requires its own extension lemma. In that result we are unable to follow this

strategy of applying Shi's theorem to extend, since the curvature bounds available
are insufficient.

3.6   Pyramid Ricci Flows

In Sect. 3.5 we stated and sketched the proof of the Local Ricci flow theorem 3.9.
This result is enough to prove Theorem 3.6 and hence settle the three-dimensional
Anderson–Cheeger–Colding–Tian conjecture, but instead of elaborating on the
details, in this section we give a different Ricci flow existence result that will not
only give the slightly stronger result Theorem 3.7, but will allow us a slightly less
technical description of the proof.

The existence result we will give is for so-called Pyramid Ricci flows. To
understand the idea, let us revisit the Local Ricci flow theorem 3.9 in the case that the
radius r is a natural number k. For each $k \in \mathbb{N}$ we obtain a Ricci flow on a parabolic
cylinder $B_{g_0}(x_0, k) \times [0, T_k]$, and in general the final existence time T_k converges
to zero as $k \to \infty$ as demonstrated by Example 3.4. The Ricci flows corresponding
to different values of k will exist on cylinders with nontrivial intersection, but it is
unreasonable to expect that they agree on these intersections.

The point of Pyramid Ricci flows is to fix this problem. Instead of obtaining a
countable number of Ricci flows on individual cylinders, indexed by k, we obtain
one Ricci flow that lives throughout the *union* of all cylinders.

Theorem 3.10 (With McLeod [23], Using [28, 29]) *Suppose* (M^3, g_0) *is a complete Riemannian manifold, and* $x_0 \in M$ *so that*

$$\begin{cases} \operatorname{Ric}_{g_0} \geq -\alpha_0 \\ \operatorname{Vol}B_{g_0}(x_0, 1) \geq v_0 > 0. \end{cases} \tag{3.20}$$

Then there exist increasing sequences $C_j \geq 1$ *and* $\alpha_j > 0$ *and a decreasing
sequence* $T_j > 0$, *all defined for* $j \in \mathbb{N}$, *and depending only on* α_0 *and* v_0, *for which
the following is true. There exists a smooth Ricci flow* $g(t)$, *defined on a subset of
spacetime*

$$\mathcal{D}_k := \bigcup_{k \in \mathbb{N}} B_{g_0}(x_0, k) \times [0, T_k],$$

satisfying that $g(0) = g_0$ *throughout* M, *and that for each* $k \in \mathbb{N}$,

$$\begin{cases} \operatorname{Ric}_{g(t)} \geq -\alpha_k \\ |\mathrm{Rm}|_{g(t)} \leq \dfrac{C_k}{t}, \end{cases} \tag{3.21}$$

throughout $\mathbb{B}_{g_0}(x_0, k) \times (0, T_k]$.

The idea of defining a Ricci flow on a subset of spacetime was used by Hochard [19], prior to this work, when he defined *partial* Ricci flows. Our flows differ in that we concede having a uniform c/t curvature bound in return for strong uniform control on the shape of the spacetime region on which the flow is defined, which is essential for the compactness arguments that use this theorem.

The theory that allows this pyramid construction revolves around a variant of the Extension lemma 3.3 known as the Pyramid Extension Lemma; see [23, Lemma 2.1] and [24, Lemma 4.1].

3.7 Proof of the 3D Anderson–Cheeger–Colding–Tian Conjecture

In this section we apply what we have discovered about Ricci flow in order to complete the programme of Sect. 3.4.4 and start the Ricci flow with a Ricci limit space in order to prove Theorems 3.6 and 3.7.

The theory can be carried out with Theorem 3.9, but here we will describe the strategy of [29] using instead Theorem 3.10, with one benefit being a slight reduction in the technical aspects of considering distance metrics on incomplete manifolds.

To reiterate the argument, we use Theorem 3.10 to flow each (M_i, g_i) generating our Ricci limit space (X, d). We call the resulting pyramid Ricci flows $g_i(t)$. The estimates on the curvature from Theorem 3.10 allow us to invoke a local version of Cheeger–Gromov–Hamilton compactness to get a type of smooth convergence

$$(M_i, g_i(t), x_i) \to (M, g(t), x_\infty) \tag{3.22}$$

to some smooth limit Ricci flow $g(t)$, which inherits the curvature bounds of the pyramid flows $g_i(t)$ and lives on a subset of spacetime similar to the domains of the pyramid flows. By keeping track of the evolution of distances, see Remark 3.5, and [29, Lemma 3.1] for the precise local statement, we obtain the following.

Theorem 3.11 (Ricci Flow From a Ricci Limit Space) *Given a Ricci limit space* $(X, d, \hat{x}_\infty)$ *that is approximated by a sequence* (M_i^3, g_i, x_i) *of complete, smooth, pointed Riemannian three-manifolds such that*

$$\begin{cases} \mathrm{Ric}_{g_i} \geq -\alpha_0 \\ \mathrm{VolB}_{g_i}(x_i, 1) \geq v_0 > 0, \end{cases} \tag{3.23}$$

there exist increasing sequences $C_k \geq 1$ *and* $\alpha_k > 0$ *and a decreasing sequence* $T_k > 0$, *all defined for* $k \in \mathbb{N}$, *and depending only on* α_0 *and* v_0, *for which the following holds.*

There exist a smooth three-manifold M, a point $x_\infty \in M$, and a complete distance metric $d_0 : M \times M \to [0, \infty)$ generating the same topology as we already have on M, such that (M, d_0, x_∞) is isometric to the Ricci limit space $(X, d, \hat{x}_\infty)$.

Moreover, there exists a smooth Ricci flow $g(t)$ defined on a subset of spacetime $M \times (0, \infty)$ that contains $\mathbb{B}_{d_0}(x_0, k) \times (0, T_k]$ for each $k \in \mathbb{N}$, with $d_{g(t)} \to d_0$ locally uniformly on M as $t \downarrow 0$. For any $k \in \mathbb{N}$ we have

$$
\begin{cases}
\text{Ric}_{g(t)} \geq -\alpha_k \\[2mm]
|\text{Rm}|_{g(t)} \leq \dfrac{C_k}{t}
\end{cases}
\tag{3.24}
$$

throughout $\mathbb{B}_{d_0}(x_0, k) \times (0, T_k]$. Finally, for all $\Omega \subset\subset M$, there exists $\sigma > 1$ depending on Ω, α_0 and v_0 such that for $t \in (0, \min\{\frac{1}{\sigma}, T\}]$ and for all $x, y \in \Omega$, we have

$$
\frac{1}{\sigma} d_{g(t)}(x, y) \leq d_{g_0}(x, y) \leq \sigma [d_{g(t)}(x, y)]^{\frac{1}{\sigma}}.
\tag{3.25}
$$

Of course, the existence of the distance metric d_0 and its relation to the original Ricci limit space is established by virtue of having the Ricci flow $g(t)$, not the other way round as the phrasing above may suggest.

The simplified distance estimate (3.25) tells us immediately that for sufficiently small $t \in (0, T]$, the identity map $id : (\Omega, d_{g(t)}) \to (\Omega, d_{g_0})$ is Hölder and $id : (\Omega, d_{g_0}) \to (\Omega, d_{g(t)})$ is Lipschitz.

This is a very rough outline. See [23, 29] for further details, and [24] for a discussion of subsequent developments and simplifications.

3.8 An Open Problem

Despite all the progress in this general area, there are some very basic problems that remain open. An example of an intriguing long-standing open problem at the heart of the matter, which contrasts with Example 3.4, is the following.

Conjecture 3.2 Given a complete Riemannian three-manifold with Ric ≥ 0, there exists a smooth complete Ricci flow continuation.

Appendix: Gromov Compactness

When defining Ricci limit spaces we used the following elementary but important compactness result.

Theorem 3.12 (Gromov Compactness, Special Case) *Given a sequence of complete pointed Riemannian n-manifolds (M_i, g_i, x_i) with a uniform lower Ricci curvature bound, we may pass to a subsequence to obtain pointed Gromov–Hausdorff convergence*

$$(M_i^n, g_i, x_i) \to (X, d, x_\infty),$$

to some pointed limit metric space (X, d, x_∞).

To illustrate the (simple) proof, let's see how one can prove the even simpler statement that for any $r > \varepsilon > 0$, there exists a metric space (X, d, x_∞) such that after passing to a subsequence, all elements of our sequence (M_i^n, g_i, x_i) are ε-close to (X, d, x_∞) in the sense that there exist maps $f_i : B_{g_i}(x_i, r) \to X$ such that

1. $f_i(x_i) = x_\infty$,
2. for all $x, y \in B_{d_i}(x_i, r)$ we have

$$|d(f_i(x), f_i(y)) - d_i(x, y)| < \varepsilon,$$

3. the ε-neighbourhood of the image $f_i(B_{d_i}(x_i, r))$ contains the ball $B_d(x_\infty, r - \varepsilon) \subset X$.

To see this, for each fixed $i \in \mathbb{N}$ pick a maximal set of points $x_i =: p_i^1, p_i^2, \ldots, p_i^m$ in $B_{g_i}(x_i, r)$ such that the balls $B_{g_i}(p_i^j, \varepsilon/9)$ lie in $B_{g_i}(x_i, r)$ and are pairwise disjoint. Maximal means that we cannot adjust the points to squeeze in even one more point with this disjointness property. That way, we can be sure that the balls $B_{g_i}(p_i^j, \varepsilon/3)$ cover all of $B_{g_i}(x_i, r)$.

The claim is that the number of such points is necessarily bounded independently of i. Assuming this is true for the moment, we can pass to a subsequence so that for each i we have the same number m of these points. The distance between each fixed pair of points is bounded above by $2r$, and below by $2\varepsilon/9$, so by passing to enough subsequences (but finitely many) we have convergence of the distances

$$d_i(p_i^j, p_i^k) \to d^{jk} > 0$$

as $i \to \infty$, where $j \neq k$. The metric space X can be chosen to be a finite collection of points $X = \{q^1, \ldots, q^m\}$, with distance

$$d(q^j, q^k) := d^{jk},$$

and we can choose $x_\infty := q^1$ and a suitable map f_i such that $f_i(p_i^j) = q^j$, giving

$$|d(f_i(p_i^j), f_i(p_i^k)) - d_i(p_i^j, p_i^k)| = |d(q^j, q^k) - d_i(p_i^j, p_i^k)| \to 0,$$

as required.

To show that the number of points $\{p_i^j\}$ is bounded independently of i, we use Bishop–Gromov and consider volumes. On the one hand, the lower Ricci bound and volume comparison (cf. Theorem 3.4 and Remark 3.2) tell us that the volume of $B_{g_i}(x_i, r)$ is uniformly bounded above independently of i. On the other hand, an argument similar to that giving (3.5) (but extended to handle lower Ricci bounds other than zero) tells us that the volume of each pairwise disjoint ball $B_{g_i}(p_i^j, \varepsilon/9)$ is uniformly bounded below by some i-independent positive number. Combining these two facts gives an i-independent upper bound on the total number of such balls that can be pairwise disjoint and fit within $B_{g_i}(x_i, r)$.

For further aspects of the proof, see for example [4, Chapter 3].

References

1. D. Burago, Y. Burago, S. Ivanov, in *A Course in Metric Geometry*. Graduate Studies in Mathematics, vol. 33 (AMS, Providence, 2001)
2. E. Cabezas-Rivas, B. Wilking, How to produce a Ricci flow via Cheeger–Gromoll exhaustion. J. Eur. Math. Soc. **17**, 3153–3194 (2015)
3. R. Bamler, E. Cabezas-Rivas, B. Wilking, The Ricci flow under almost non-negative curvature conditions. Inventiones **217** 95–126
4. J. Cheeger, *Degeneration of Riemannian Metrics under Ricci Curvature Bounds* (Edizioni della Normale, 2001)
5. J. Cheeger, T.H. Colding, On the structure of spaces with Ricci curvature bounded below I. J. Differ. Geom. **45**, 406–480 (1997)
6. B.-L. Chen, X.-P. Zhu, Uniqueness of the Ricci flow on complete noncompact manifolds. J. Differ. Geom. **74**, 119–154 (2006)
7. B.-L. Chen, Strong uniqueness of the Ricci flow. J. Differ. Geom. **82**, 363–382 (2009)
8. B. Chow, The Ricci flow on the 2-sphere. J. Differ. Geom. **33**, 325–334 (1991)
9. B. Chow, D. Knopf, in *The Ricci Flow: An Introduction*. Mathematical Surveys and Monographs, vol. 110 (American Mathematical Society, Providence, 2004)
10. P. Daskalopoulos, M. del Pino, On a singular diffusion equation. Commun. Anal. Geom. **3**, 523–542 (1995)
11. E. DiBenedetto, D. Diller, About a singular parabolic equation arising in thin film dynamics and in the Ricci flow for complete $\mathbb{R}^2$, in *Partial Differential Equations and Applications: Collected Papers in Honor of Carlo Pucci*. Lecture Notes in Pure and Applied Mathematics, vol. 177 (CRC Press, Boca Raton, 1996), pp. 103–119.
12. P. Gianniotis, The Ricci flow on manifolds with boundary. J. Differ. Geom. **104**, 291–324 (2016)
13. G. Giesen, P.M. Topping, Existence of Ricci flows of incomplete surfaces. Commun. Partial Differ. Equ. **36**, 1860–1880 (2011). https://arxiv.org/abs/1007.3146
14. G. Giesen, P.M. Topping, Ricci flows with unbounded curvature. Math. Zeit. **273**, 449–460 (2013). https://arxiv.org/abs/1106.2493
15. G. Giesen, P.M. Topping, Ricci flows with bursts of unbounded curvature. Commun. Partial Differ. Equ. **41**, 854–876 (2016). https://arxiv.org/abs/1302.5686
16. R.S. Hamilton, Three-manifolds with positive Ricci curvature. J. Differ. Geom. **17**, 255–306 (1982)
17. R.S. Hamilton, The Ricci flow on surfaces, in *Mathematics and General Relativity (Santa Cruz, 1986)*. Contemporary Mathematics, vol. 71 (American Mathematical Society, Providence, 1988), pp. 237–262

18. R.S. Hamilton, The formation of singularities in the Ricci flow, in *Surveys in Differential Geometry*, vol. II (Cambridge, MA, 1993) (International Press, Cambridge, 1995), pp. 7–136
19. R. Hochard, Short-time existence of the Ricci flow on complete, non-collapsed 3-manifolds with Ricci curvature bounded from below. http://arxiv.org/abs/1603.08726v1
20. C. Ketterer, Cones over metric measure spaces and the maximal diameter theorem. J. Math. Pures Appl. **103**, 1228–1275 (2015)
21. B. Kotschwar, An energy approach to the problem of uniqueness for the Ricci flow. Commun. Anal. Geom. **22**, 149–176 (2014)
22. Y. Lai, Ricci flow under local almost non-negative curvature conditions. Adv. Math. **343**, 353–392 (2019)
23. A.D. McLeod, P.M. Topping, Global regularity of three-dimensional Ricci limit spaces. Trans. Am Math. Soc. Ser. B (2018). https://arxiv.org/abs/1803.00414
24. A.D. McLeod, P.M. Topping, Pyramid Ricci flow in higher dimensions. Math. Z. (2020). https://doi.org/10.1007/s00209-020-02472-1
25. G. Perelman, The entropy formula for the Ricci flow and its geometric applications (2002). http://arXiv.org/abs/math/0211159v1
26. W.-X. Shi, Deforming the metric on complete Riemannian manifolds. J. Differ. Geom. **30**, 223–301 (1989)
27. M. Simon, Ricci flow of non-collapsed three manifolds whose Ricci curvature is bounded from below. J. Reine Angew. Math. **662**, 59–94 (2012)
28. M. Simon, P.M. Topping, Local control on the geometry in 3D Ricci flow. https://arxiv.org/abs/1611.06137
29. M. Simon, P.M. Topping, Local mollification of Riemannian metrics using Ricci flow, and Ricci limit spaces. Geom. Topol. (2017). https://arxiv.org/abs/1706.09490
30. P.M. Topping, *Lectures on the Ricci flow*. London Mathematical Society Lecture Note Series), vol. 325 (Cambridge University Press, Cambridge, 2006). https://www.warwick.ac.uk/~maseq/RFnotes.html
31. P.M. Topping, Ricci flow compactness via pseudolocality, and flows with incomplete initial metrics. J. Eur. Math. Soc. **12**, 1429–1451 (2010)
32. P.M. Topping, Uniqueness and nonuniqueness for Ricci flow on surfaces: reverse cusp singularities. Int. Math. Res. Not. **2012**, 2356–2376 (2012). https://arxiv.org/abs/1010.2795
33. P.M. Topping, Uniqueness of instantaneously complete Ricci flows. Geom. Topol. **19**(3), 1477–1492 (2015). https://arxiv.org/abs/1305.1905
34. P.M. Topping, H. Yin, Sharp decay estimates for the logarithmic fast diffusion equation and the Ricci flow on surfaces. Ann. PDE **3**, 6 (2017). https://doi.org/10.1007/s40818-017-0024-x
35. P.M. Topping, H. Yin, Rate of curvature decay for the contracting cusp Ricci flow. Commun. Anal. Geom. **28** (2020). https://www.intlpress.com/site/pub/pages/journals/items/cag/_home/acceptedpapers/index.php
36. J.-L. Vázquez, in *Smoothing and Decay Estimates for Nonlinear Diffusion Equations*. Oxford Lecture Series in Mathematics and Its Applications, vol. 33 (2006)
37. C. Villani, Synthetic theory of Ricci curvature bounds. Jpn. J. Math. **11**, 219–263 (2016)
38. D. Yang, Convergence of Riemannian manifolds with integral bounds on curvature. I. Ann. Sci. École Norm. Sup. **25**, 77–105 (1992)

Chapter 4
Pseudo-Hermitian Geometry in 3D

Paul C. Yang

Abstract This chapter concerns CR geometry, a research field for which there is an extremely fruitful interaction of different ideas, ranging from Differential Geometry, Partial Differential Equations and Complex Analysis. First, some basic concepts of the subject are introduced, as well as some conformally covariant operators. Then some surprising relations are shown between the embeddability of abstract CR structures, the spectral properties of the Paneitz operator and the attainment of the Yamabe quotient. Finally, some surface theory is treated, in relation to the isoperimetic problem, the prescribed mean curvature problem and to some Willmore-type functionals.

4.1 Introduction

In this series of lectures I plan to discuss several geometric/analytic problems in CR-geometry. They are, in order of presentation: the embedding problem, the CR Yamabe problem, the Q-prime curvature equation and finally the geometry of surfaces in the Heisenberg group. After a brief introduction to the local invariants defined by Tanaka and Webster we will follow the following table of contents. I would like to thank Matt Gursky and Andrea Malchiodi for the invitation to give this series of lectures at the Cetraro Summer School.

1. Local invariants: the Webster connection, torsion and curvature.
2. Conformally covariant operators, the CR conformal Laplacian, the CR Paneitz operator and the Q-curvature.
3. Pluriharmonic functions, Fefferman equations, psuedo-Einstein contact form, the P-prime operator and the Q-prime curvature, the Burns-Epstein invariant.

P. C. Yang (✉)
Department of Mathematics, Princeton University, Princeton, NJ, USA
e-mail: yang@math.princeton.edu

© The Editor(s) (if applicable) and The Author(s), under exclusive licence to Springer Nature Switzerland AG 2020
M. J. Gursky, A. Malchiodi (eds.), *Geometric Analysis*, Lecture Notes in Mathematics 2263, https://doi.org/10.1007/978-3-030-53725-8_4

4. The embedding problem and the Kohn Laplacian, sign of the Paneitz operator.
5. Positive mass theorem.
6. The Q-prime curvature equation: a sharp inequality for the total Q-prime curvature, an existance theorem for the Q-prime curvature.
7. Isoperimetric inequality for the total Q-prime curvature.
8. Geometry of surfaces in the Heisenberg group, the mean curvature and the angle function α, the analysis of the Codazzi equation.
9. Analogues of the Willmore functional, renormalized area, examples and open problems.

In these notes we will follow the notational conventions in [26].

4.2 Definitions

A pseudo-Hermitian 3-manifold is a triple (M, θ, J) where θ is a 1-form satisfying $\theta \wedge d\theta \neq 0$ and letting $\xi = Ker J$, the almost complex structure is given by $J : \xi \to \xi$ satisfying $J^2 = -I$. The Reeb vector field T is uniquely determined by $\theta(T) = 1$ and $d\theta(T, \cdot) = 0$.

Let H denote the $+i$ eigenspace of $\mathbb{C} \otimes \xi$ and Z_1 is a local frame for H normalized by the condition $d\theta(Z_1, \bar{Z}_1) = 1$. Let $\{\theta, \theta^1, \theta^T\}$ be the dual 1-form to $\{T, Z_1, Z_{\bar{1}}\}$.

$$d\theta = i\theta^1 \wedge a^{\bar{1}}.$$

The Webster connection is characterized by

$$\nabla Z_1 = \omega_1^1 \otimes Z_1 \qquad \nabla Z_{\bar{1}} = \omega_{\bar{1}}^{\bar{1}} \otimes Z_{\bar{1}}, \nabla T = 0,$$

where ω_1^1 is the 1-form uniquely determined by

$$d\theta^1 = \theta^1 \wedge \omega_1^1 + A_{\bar{1}}^1 \theta \wedge \theta^{\bar{1}}, \omega_1^1 + \omega_{\bar{1}}^{\bar{1}} = 0.$$

$A_{\bar{1}}^1$ is called the *torsion tensor*.

$$d\omega_1^1 = R\theta^1 \wedge \theta^{\bar{1}} + A_{1,\bar{1}}^{\bar{1}}\theta^1 \wedge \theta - A_{\bar{1},1}^1 \theta^{\bar{1}} \wedge \theta$$

where R is the *Webster scalar curvature*.

Examples

1. The Heisenberg $\mathbb{H}^1 = \{(x, y, t) \in \mathbb{R}^3 | \theta = dt + xdy - ydx\}$ $Z_1 = \partial_x - i\partial_y + (ix + y)\partial_t$, $T = \partial_t$. This is the flat model: $A = 0$ and $R = 0$.
2. The unit sphere $S^3 = \{(z_1, z_2) \in \mathbb{C}^2 | |z_1|^2 + |z_2|^2 = 1\}$

$$\theta = i \sum_{k=1}^{2} (\bar{z}^k dz^k - z^k d\bar{z}^k), \ z_1 = \bar{z}^2 \partial_{z^1} - \bar{z}^1 \partial_{z^2}$$

$A = 0$ and $R = 1$.

The Cayley transform: $F : S^3 \backslash \{(0, -1)\} \rightarrow H^1$

$$x = Re \frac{z_1}{1 + z_2}, \ y = Im \frac{z_1}{1 + z_2}, \ t = \frac{1}{2} Re \left\{ i \frac{1 - z_2}{1 + z_2} \right\}.$$

It is elementary to observe that the nondegeneracy condition $\theta \wedge d\theta \neq 0$ means the distribution $\xi = Ker\theta$ is not integrable. That is if X and Y are vector fields satisfying $\theta(X) = \theta(Y) = 0$ then $0 \neq d\theta(X, Y) = X\theta(Y) - Y\theta(X) - \theta([X, Y])$ if X, Y are independent.

The basic regularity theory of the $\bar{\partial}_b$ equation on the Heisenberg group is established by Folland and Stein [21] making use of Fourier analysis on the Heisenberg group. Explicit kernels are given. In this way they recover the basic estimates of Kohn: $C||u||_{\frac{1}{2}} \leq ||\bar{\partial}_b u||_{L^2} + ||u||_{L^2}$.

By introducing the space $S_k^p = \{$ function with k tangential derivatives in $L^p\}$, they obtained the more precise result:

Let $L_\alpha u = f$ where $L_\alpha u = \Delta_b u - i\alpha T$ when $\pm\alpha \neq n, n + 2, n_x, \ldots$ then the following local estimates hold if $f \in S_k^p(U)$ and $V \subset\subset U$ then $u \in S_{k+2}^p(V)$.

Definition A CR structure is locally spherical if there is a local biholomorphic map to the standard sphere.

Definition The Cartan tensor [18] is defined as

$$Q_J = iQ_{\bar{1}}^{\bar{1}}\theta^1 \otimes Z\bar{1} - iQ_1^{\bar{1}}\theta^{\bar{1}} \otimes Z_1$$

where

$$Q_1^{\bar{1}} = \frac{1}{6}R_{,1}^{\bar{1}} + \frac{i}{2}RA_1^{\bar{1}} - A_{1,0}^{\bar{1}} - \frac{2i}{3}A_{1,T}^{\bar{1}\bar{1}}.$$

The vanishing of Q_J is necessary and sufficient for the structure to be locally spherical.

4.3 Conformally Covariant Operators

The CR conformal Laplacian is given by:

$$L_\theta = -4\Delta_b + R \text{ where } \Delta_b u = u_{,1}{}^1 + u_{,\bar{1}}^{\bar{1}}$$

It is elementary to check that under a conformal change of contact form $\tilde{\theta} = u^2\theta$

$$L_{\tilde{\theta}} f = u^{-3} L_\theta(uf)$$

and

$$L_\theta u = \tilde{R} u^3$$

where $\tilde{R}$ is the scalar curvature of $\tilde{\theta}$. Just like the Yamabe problem, there is a variational functional

$$q[u] = \frac{\int Lu \cdot u\theta \wedge d\theta}{\|u\|_4^2}.$$

The Euler equation for critical points of $q[u]$ is the CR Yamabe equation

$$L_\theta u = \lambda u^3$$

where λ is the Lagrange multiplier.

The CR Paneitz operator is given by

$$P\varphi = 4(\varphi_{1\bar{1}}^{\bar{1}} + i A_{11}\varphi')^1 = (P_3\varphi)^1.$$

The operator P is real, self-adjoint operator and satisfies the covariance property: for $\tilde{\theta} = u^2\theta$

$$P_{\tilde{\theta}}\varphi = u^{-4} P_\theta\varphi.$$

The third order operator P_3 characterizes pluriharmonic functions [26]. The Paneitz operator is closely related to *Kohn's Laplacian*

$$\Box_b = -\Delta_b + iT :$$

we have

$$P\phi = \frac{1}{4}(\Box_b \circ \bar{\Box}_b\varphi - 4i(A^{11}\varphi_1)_1)$$

$$= \frac{1}{4}(\bar{\Box} \circ \Box_b\varphi + 4i(A^{\bar{1}\bar{1}}\varphi_{\bar{1}})_{\bar{1}})$$

When the torsion vanishes, the operators $\Box_b$ and $\bar{\Box}_b$ commute, hence they are simultaneously diagonalizable, and in this case it follows that eigenvalues of P are non-negative. The Paneitz operator is closely connected with Q curvature

$$Q = -\triangle_b R - 2Im A_{11}^{11},$$

which appears [23] as the coefficient of the logarithm term in the asymptotic expansion of the Szego kernel for a C^∞ strictly pseudoconvex domain $\Omega \in \mathbb{C}^2$ with respect to a volume element $\theta \wedge d\theta$ on $\partial\Omega$

$$S(z, \bar{z}) = \varphi(z)\rho(z)^{-2} + \psi(z)\log\rho(z)$$

with $\varphi, \psi, \in C^\infty(\bar{\Omega})$ where ρ is a defining function $\rho > 0$ on Ω. Under change of contact form $\tilde{\theta} = e^{2f}\theta$, $\psi_0 = \psi|_{\partial\Omega}$ transforms as

$$\tilde{\psi}_0 = e^{-4f}\left(\psi_0 + \frac{1}{12}Pf\right).$$

Closely related is the 1-form

$$W_1\theta^1 = (R_{,1} - iA_{11}^1)\theta^1,$$

so that

$$W_{1,}^1 = \frac{1}{2}\triangle_b R + Im A_{11}^{11},$$

and W_1 transforms as

$$\tilde{W}_1 = e^{-3f}(W_1 - 6P_3 f)$$

where $P_3 f = f_{11}^{\bar{1}} + iA_{11}f^1$ is the operator that defines pluriharmonic functions.

4.4   Fefferman's Equation, Pseudo-Einstein Contact Form, P' and Q'

The question whether there is a non-zero coefficient ψ_0 in the Szego kernel expansion is answered by the work of Fefferman [20] where he considered the following Monge Ampere equation for a defining function u of a strictly pseudo convex domain Ω in $\mathbb{C}^2$:

$$J[u] = \det \begin{bmatrix} u & \dfrac{\partial u}{\partial \bar{z}_j} \\ \dfrac{\partial u}{\partial z_i} & \dfrac{\partial^2 u}{\partial z_i \partial \bar{z}_j} \end{bmatrix} = 1 \qquad \text{on } \bar{\Omega}. \qquad (4.4.1)$$

A smooth solution of this equation in the interior [17] gives a complete Kahler–Einstein metric g whose Kahler form is $i\partial\bar{\partial}\log(u)$. In general, the solution may not be smooth at the boundary. C. Fefferman provides an iterative process to compute approximate solutions to (4.4.1). Let ψ be any smooth defining function in $\partial\Omega$, let

$$u_1 = \frac{\psi}{J[\psi]^{1/3}}$$

$$u_s = u_{s-1}(1 + \frac{1 - J[u_{s-1}]}{(n + 2 - s)s}) \qquad 2 \le s \le n + 1$$

Then u_s satisfies $J[u_s] = 1 + O(\psi^s)$.

It turns out that for the contact form $\theta = Im\bar{\partial}u_3$ satisfies the equation $W_1 = 0$ and hence has $Q = 0$.

Contact forms θ satisfying the condition $R_{,1} - iA^1_{11,} = 0$ are called pseudo-Einstein on account of the following consequence of the Bianchi identity in M^{2n+1}

$$\nabla^{\bar{\beta}}\left(R_{\alpha\bar{\beta}} - \frac{1}{n}Rh_{\alpha\bar{\beta}}\right) = \frac{n-1}{n}\left(\nabla_\alpha R - in\nabla^\beta A_{\alpha\beta}\right).$$

Thus when $n = 1$, the pseudo-Einstein condition is the "residue in n" of the general pseudo-Einstein condition.

The pseudo-Einstein contact forms are in $1-$to-1 correspondence with pluriharmonic functions according to

Lemma *If θ is pseudo-Einstein then $\tilde{\theta} = e^{2f}\theta$ is pseudo-Einstein iff f is pluriharmonic.*

This is evident from the transformation rule for W_1 at the end of Sect. 4.3.

P-prime operator and Q-prime curvature:

In general dimensions there is the CR-Paneitz operator of order 4 on (M^{2n+1}, θ, τ) (Gover/Graham)

$$P_4 f = \Delta_b^2 f + \nabla_0^2 f - 4Im\nabla^\alpha(A_{\alpha\beta}\nabla^\beta f) + 4Re(\nabla_{\bar{\beta}}P^{\alpha\bar{\beta}}\nabla_\alpha)$$

$$- 4\frac{n^2 - 1}{n}Re(\nabla^\beta(P\nabla_\beta f)) + \frac{n-1}{2}Q_4 f$$

where

$$Q_4 = \frac{2(n+1)^2}{n(n+2)}\Delta_b P - \frac{4}{n(n+1)}Im(\nabla^\alpha\nabla^\beta A_{\alpha\beta}) - 2\frac{(n-1)}{n}|A_{\alpha\beta}|^2$$

$$- \frac{2(n+1)}{n}|\dot{P}_{\alpha\bar{\beta}}|^2 + \frac{2(n-1)(n+1)^2}{n^2}P^2$$

where

$$P_{\alpha\bar{\beta}} = \frac{1}{n+2}(R_{\alpha\bar{\beta}} - \frac{1}{2(n+1)}Rh_{\alpha\bar{\beta}})$$

$$P = tr P = \frac{R}{2(n+1)}$$

P_4 is conf. covariant of degree $(-\frac{n-1}{2}, \frac{n+3}{2})$.

Definition ([7]) $P_4' f = \frac{2}{n-1}P_4 f$ for $n > 1$, and $P_4' f = \lim_{n \to 1} \frac{2}{n-1}P_4 f$ for $f \in \mathcal{P}$ when $n = 1$.

Observe

$$P_4' f = 4\Delta_b^2 f - 8 Im \nabla^\alpha (\Delta_{\alpha\beta} \nabla^\beta f) - 4 Re(\nabla^\alpha R \nabla_\alpha f)$$

$$+ \frac{8}{3} Re((\nabla_\alpha R - i\nabla^\beta A_{\alpha\beta})\nabla^\alpha f) + \frac{2}{3}(-\Delta_b R - 2\nabla^\alpha \nabla^\beta A_{\alpha\beta})f$$

So

$$P_4'(1) = Q$$

$$Q_4' = \frac{2}{n-1}P'(1) = \frac{4}{(n-1)^2}P_4(1)$$

$$= -2\Delta_b R + R^2 - 4|A_{\alpha\beta}|^2. \tag{4.4.2}$$

The total Q'-curvature and the Burns–Epstein invariant
   In case $M^3 = \partial X$ where X is a strictly pseudoconvex domain in $\mathbb{C}^2$ the solution of Fefferman's equation gives X a complete Kahler Einstein metric and there is a Gauss Bonnet type formula

$$C\chi(X) = \int_X (c_2 - \frac{1}{3}c_1^2) + \oint_M Q'\theta \wedge d\theta$$

where Q' is the Q' curvature, and $\oint Q'\theta \wedge d\theta$ agrees with the Burns–Epstein invariant [4] given by the integral of the 3-form

$$i\left[\left(-\frac{2i}{3}\omega_1^1 \wedge d\omega_1^1 + \frac{1}{6}(R\theta) \wedge d\omega_1^1 - 2|A|^2\theta \wedge d\theta\right)\right] \tag{4.4.3}$$

where ω_1^1 is the connection form, R the scalar curvature, A the torsion.

Lemma *A contact form θ is pseudo-Einstein iff there exists frames θ, θ^1 so that $h_{\alpha\beta} = \delta_{\alpha\bar{\beta}}$ and $\omega_1^1 + (\frac{i}{n})R\theta = 0$*

Proof This is elementary.

It then follows that (4.4.3) yields (4.4.2).

Remark For KE metrics the integrand $c_2 - \frac{1}{3}c_1^2$ is a sum of squares, and its integral is always finite. No renormalization is needed.

4.5   The Embedding Problem

The almost complex structure J gives a splitting

$$d_b = \partial_b + \bar{\partial}_b,$$

where $\partial_b f = Z_1(f)\theta^1, \bar{\partial}_b f = Z_{\bar{1}}(f)\theta^{\bar{1}}$

The usual integrability condition

$$\bar{\partial}_b^2 = 0$$

becomes vacuous in dimension $2n + 1 = 3$. In dimensions $2n + 1 \geq 5$, under this integrability condition, closed CR manifolds may be embedded holomorphically in $\mathbb{C}^N$ [2]. Indeed Kohn showed that if $\Box_b$ on functions has closed range, then the CR structure is embeddable.

Theorem ([6]) *If (M^3, θ, J) satisfy the condition $P \geq 0$ and the Webster scalar curvature $R \geq c > 0$, then the non-zero eigenvalues λ of $\Box_b$ have the lower bound*

$$\lambda \geq \min R \geq c > 0.$$

It follows that $\Box_b$ has closed range. Hence the CR structure is embeddable.

The proof is a simple application of the following Bochner formula

$$-\frac{1}{2}\Box_b|\bar{\partial}_b\varphi|^2 = (\varphi_{\bar{1}\bar{1}}\bar{\varphi}_{11} + \varphi_{\bar{1}1}\bar{\varphi}_{1\bar{1}}) - \frac{1}{2}\langle\bar{\partial}_b\varphi, \bar{\partial}_b\Box_b\varphi\rangle - \langle\bar{\partial}_b\Box_b\varphi, \bar{\partial}_b\varphi\rangle$$
$$- \langle\bar{P}_3\varphi, \bar{\partial}_b\varphi\rangle + R|\bar{\partial}_b\varphi|^2.$$

Apply this to an eigenfunction φ with non-zero eigenvalue. Integrating this Bochner formula we find

$$0 = \int \varphi_{\bar{1}\bar{1}}\bar{\varphi}_{11} + \int \varphi_{\bar{1}1}\bar{\varphi}_{1\bar{1}} - \frac{3}{2}\lambda \int |\bar{\partial}_b\varphi|^2 + \int \langle P\varphi, \varphi\rangle + \int R|\bar{\partial}_b\varphi|^2.$$

Rewriting

$$\int \varphi_{1\bar{1}}\bar{\varphi}_{1\bar{1}} = \frac{1}{4}\int \langle \Box, \varphi, \Box_b\varphi\rangle = \frac{\lambda}{2}\int |\bar{\partial}_b\varphi|^2,$$

we obtain

$$\lambda \int |\bar{\partial}_b\varphi|^2 = \int |\varphi_{1\bar{1}}|^2 + \int P\varphi \cdot \bar{\varphi} + \int R|\bar{\partial}_b\varphi|^2 \geq \int P\varphi \cdot \bar{\varphi} + \int R|\bar{\partial}_b\varphi|^2.$$

Example of Rossi [3]

On (S^3, θ_0, J_0) let $Z_1 = \bar{z}_2\partial_{z_1} - \bar{z}_1\partial_{z_2}$.

Change the CR structure to

$$Z_{1(t)} = Z_1 + t\bar{Z}_1$$

where $|t| < 1$, keeping the same contract form.

An elementary argument using spherical harmonics shows that all holomorphic functions with respect to $\bar{Z}_{1(t)}f = 0$ are even, i.e. $f(z_1, z_2) = f(-z_1, -z_2)$. Hence it is not possible to separate points by holomorphic functions.

However, a relatively easy computation shows

$$R(t) = \frac{2(1 + t^2)}{1 - t^2} \geq 2.$$

On the other hand, $(S^3/\mathbb{Z}_2, \theta_\lambda J_t)$ does embed [10].

Kohn's solution of $\Box_b$ on (M^3, θ, J) if $M^3 = \partial\Omega$ in a Stein manifold. There exists an solution operator K on $L^2(M^3)$

$$\Box_b \circ K = K \circ \Box_b = Id - S$$

where S is the Szego projection with respect to $\theta \wedge d\theta$. Such K exists when $\bar{\partial}_b : L^2 \to L^2_{(0,1)}$ has closed range and $||u||_{1/2} \leq C||\bar{\partial}_b u||^2$ if $u \perp N(\bar{\partial}_b)$.

When Is $P \geq 0$?

Recall $P\varphi = \delta_b P_3\varphi$ where $P_3\varphi = (\varphi^{\bar{1}}_{11} + iA_{11}\varphi^1)\theta^1$ and $P_3\varphi = 0$ is equivalent to φ being pluriharmonic written $\varphi \in \mathcal{P}$. It follows that $Ker P \subset \mathcal{P}$.

Typically $\mathcal{P}$ is an infinite dimensional space if (M^3, θ, J) is embeddable.

Let us write $Ker P = \mathcal{P} \oplus W$.

For an embeddable CR structure, W is at most finite dimensional.

When Is W Empty?

About the supplementary space W the following estimate is useful:

Lemma *Suppose* $\lambda_1(\Box_b) \geq C > 0$, *for* $f \in S^{4,2} \cap \mathcal{P}^\perp$ *there exists* C_1:

$$C_1 \|f\|_{S^{4,2}} \leq \|Pf\|_2 + \|f\|_2.$$

Proof $f \in \mathcal{P}^\perp$ implies $f \perp Ker \, \partial_b$. Hence there exists solution ψ to $\bar{\Box}_b \psi = f$.
Let h be an antiholomorphic function

$$\langle \Box_b f, h \rangle = \langle \Box_b \bar{\Box}_b \psi, h \rangle$$

$$= \langle \psi, (\bar{\Box}_b \Box_b - 2\bar{Q})h \rangle + 2\langle \psi, \bar{Q}h \rangle,$$

where

$$P = \bar{\Box}_b \Box_b - 2\bar{Q},$$

therefore

$$\langle \Box_b f - 2Q\psi, h \rangle = 0.$$

Also

$$\lambda_1(\Box_b) \geq c > 0 \Rightarrow$$

$$C_1 \|f\|_{S^{4,2}} \leq \|\Box_b f\|_{S^{2,2}} + \|f\|_2$$

$$\leq \|\Box_b f - 2Q\psi\|_{S^{2,2}} + 2\|Q\psi\|_{S^{2,2}} + \|f\|_2. \tag{*}$$

Since $(\Box f - 2Q\psi) \perp$ antiholomorphic

$$C_2 \|\Box_b f - 2Q\psi\|_{S^{2,2}} \leq \|\bar{\Box}_b(\Box_b f - 2Q\psi)\|_2 + \|\Box_b f - 2Q\psi\|_2 \tag{**}$$

$$\Rightarrow C_3 \|f\|_{S^{4,2}} \leq \|Pf\|_2 + \|\bar{Q}f\|_2 + \|Q\psi\|_{S^{2,2}} + \|f\|_{S^{2,2}} + \|f\|_2$$

$$\Rightarrow C \|\psi\|_{S^{4,2}} \leq \|f\|_{S^{2,2}} + \|f\|_2$$

$$\Rightarrow C_3 \|f\|_{S^{4,2}} \leq \|Pf\|_2 + \underbrace{\|f\|_{S^{2,2}}}_{\leq} + \|f\|_2$$

$$\times \epsilon \|f\|_{S^{4,2}} + C_\epsilon \|f\|_2. \qquad \Box$$

Corollary *For an embedded CR structure* $Ker \, P = \mathcal{P} + W$, *where* $\dim W < \infty$

Definition For a one-parameter-family of CR structures (M^3, θ, J_t) we say that $\mathcal{P}^t$
is stable if for any $\varphi \in \mathcal{P}^t$, and $\epsilon > 0$, there exists $\delta > 0$ so that for $|t - s| < \delta$ there
is a $f_s \in \mathcal{P}^s$ so that

$$\|\varphi - f_s\|_2 < \epsilon.$$

Theorem ([9]) *Let (M^3, θ, J^t) be a family of embedded CR structures for $t \in [-1, 1]$ with the following*

1. *J^t is real analytic in the parameter t*
2. *The Szego projection $S^t = F^{2,0} \rightarrow Ker\,\bar{\partial}_b^{\,t} \subset F^{2,0}$ vary continuously in the parameter t (def $F^{2,0}$ later)*
3. *For J^0 we have $P^0 \geq 0$ and $Ker\,P^0 = \mathcal{P}^0$*
4. *There exists a uniform $c > 0$ s.t.*

$$\min_{-1 \leq t \leq 1} R^t \geq c > 0$$

5. *The pluriharmonics $\mathcal{P}^t$ are stable with respect to t*

then

$$P^t \geq 0 \text{ and } Ker\,P^t = \mathcal{P}^t \text{ for all } t \in [-1, 1].$$

Proof By continuity:
   Let

$$S = \{t \in [-1, 1] | P^t \geq 0 \text{ and } Ker\,P^t = \mathcal{P}^t\}.$$

S is open:
   The small eigenvalues of P^t are finitely many and parameterized by $\lambda_i(t)$ real analytic in t, $i = 1, \cdots, k$. To be precise let there be no eigenvalues of P^0 in the intervals $(-r, 0) \vee (0, r)$, and the eigenvalue of P^t in these intervals are called small eigenvalues of P^t, $|t|$ small. Assume to the contrary u_t is an eigenfunction for P^t with small eigenvalue. Write $u_t = u_0 + f_t$ where $u_0 \in Ker\,P^0$ and $\|f_t\|_2 = o(1)$, $\|u_0\|_2 = 1$. Stability implies that there is $g_t \in Ker\,P^t$ such that $\|u_0 - g_t\| < \epsilon$

$$\begin{aligned}
0 &= \langle u_t, g_t \rangle \\
&= \langle u_t, u_0 \rangle + \langle u_t, g_t - u_0 \rangle \\
&= 1 + \langle f_t, u_0 \rangle + \langle u_t, g_t - u_0 \rangle = 1 + o(1). \quad\quad (*)
\end{aligned}$$

Next we check $W^t = \{0\}$ for $|t|$ small. Observe the constant c in the Lemma can be taken uniformly in t. So if to the contrary, there are $f_{t_k} \in W_{t_k}$ $\|f_{t_k}\|_1 = 1$, $t_k \rightarrow t_0$ the Lemma implies $\|f_{t_u}\|_{W^{4,2}} \leq c$. Hence a subsequence converges strongly in L^2.

$$f_{t_k} \rightarrow f_0 \quad\quad \|f_0\|_{S^{4,2}} \leq c$$

$$P^{t_0} f_0 = 0 \Rightarrow f_0 \in \mathcal{P}^{t_\circ}$$

Stability implies that given $\epsilon > 0 \; \exists \delta > 0$ s.t.

$$|t - t_0| < \delta \Rightarrow \exists \psi_t \in \mathcal{P}^t$$

$$\|f_0 - \psi_t\|_2 < \epsilon$$

$$1 = \|f_0\|_2^2 = \underset{\downarrow \atop 0}{\langle f_0 - \psi_{t_k}, f_0 \rangle} + \underset{\downarrow \atop 0}{\langle \psi_{t_k}, f_0 - f_{t_k} \rangle} + \underset{\| \atop 0}{\langle \psi_{t_k}, f_{t_k} \rangle}$$

$$= o(1) \tag{*}$$

For closedness of S we need to introduce the Garfield-Lee complex. Let

$$E^{0,0} = \langle 1 \rangle, \; E^{1,0} = \langle \theta^1 \rangle, \; E^{0,1} = \langle \theta^{\bar{1}} \rangle, \; F^{2,0} = \langle \theta^1 \wedge \theta \rangle, \; F^{1,1} = \langle \theta^{\bar{1}} \wedge \theta \rangle$$

$F^{2,1} = \langle \theta \wedge d\theta \rangle$ be the line bundles

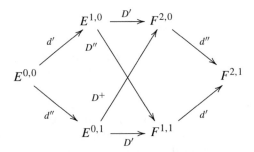

Here $d' = \partial_b, d'' = \bar{\partial}_b$ and D', D'' and D^+ are the second order operators introduced by Rumin [29]

$$D'(\sigma_1, \theta^1) = (-i\sigma_{1,\bar{1}1} - \sigma_{1,0})\theta^1 \wedge \theta$$

$$D''(\sigma_1\theta^1) = (-i\sigma_{1,\bar{1}\bar{1}} - A_{\bar{1}\bar{1}}\sigma_1)\theta^{\bar{1}} \wedge \theta$$

$$D'(\sigma_{\bar{1}}\theta^{\bar{1}}) = \langle i\sigma_{\bar{1},1\bar{1}} - \sigma_{\bar{1},0}\rangle\theta^1 \wedge \theta$$

$$D^+(\sigma_{\bar{1}}\theta^{\bar{1}}) = \langle i\sigma_{\bar{1},11} - A_{11}\sigma_{\bar{1}}\rangle\theta^1 \wedge \theta$$

It follows that

$$P_3 = -iD^+d''$$

$$P = -id''D^+d''$$

The supplementary subspace W may be realized as

$$W \cong Ker d'' \cap im D^+ d'' \subset F^{2,0}$$

$$\cong Ker d'' \cap Im P_3 \subset F^{2,0}$$

Let

$$F^t = Ker \bar{\partial}_b^t \cap Im P_3^t \subset F^{2,0},$$

so that

$$dim F^t = dim W^t.$$

Claim $dim F^t$ is a lower semicontinuous function of t.

Let S^t be the Szego projector in $F^{2,0}$. Then rank $(S_0^t P_3^t) = dim F^t < \infty$. For fixed $\varphi \in E^{00}$ $\psi \in (F^{2,0})^* \otimes L^2$,

$$h(t) = \langle S_0^t P_3^t (\varphi), \psi \rangle$$

is continuous. It suffices to check that

$$G = \{t \in [-1, 1] | \text{ rank } (S_0^t P_3^t) > a\}$$

is open for any a.
   Set $r = $ rank $S^{t_0}_0 P_3^{t_0} > a$.
   There exist $\{\varphi_i\}_{i=1}^r$ and functions $\{\psi_i\}_{i=1}^r$, s.t.

$$h_{ij}(t_0) = \langle S^{t_0} \circ P_3^{t_0} \varphi_i, \psi_j \rangle$$

is continuous:

$$\det h_{ij}(t_0) \neq 0$$

$$\Rightarrow \det h_{ij}(t) \neq 0 \qquad \text{for } |t - t_0| \text{ small}$$

To finish the proof,
   Let $t_n \in S, t_n \to t_0$. We have $P^{t_0} \geq 0$ and

$$\overline{\lim_{t_n \to t_0}} dim W^{t_n} \leq dim W^{t_0}$$

Lower semi-continuity implies then $dim W^{t_0} \leq \liminf dim W^{t_n} = 0$. □

   A typical family would start with J^0 so that the torsion vanishes. For example, the ellipsoids form such a family with J^0 the standard sphere.

In [1] Bland determined embeddable deformations of J on the standard 3-sphere in terms of conditions on the Fourier coefficients of $\dot{J}$.

In a recent preprint [30], Takeuchi showed that the embeddedness implies $P \geq 0$ and the kernel of P consist of pluriharmonics.

4.6   The Positive Mass Theorem

For (M^3, θ, J) we consider the variational problem to find the minimizer of the Sobolev quotient

$$q[u] = \frac{\int Lu \cdot u}{||u||_4^2}.$$

The interesting case is when the CR conformal Laplacian L_θ is a positive operator. The same argument as in the case of conformal geometry applies to reduce this problem to finding a test function u for which $q[u]$ is strictly smaller than the corresponding constant as the standard 3-sphere

$$q_0[1] = \frac{\int (L_0 \cdot 1 \cdot) \cdot 1}{(\text{vol } S^3)^{\frac{1}{2}}}.$$

It turns out that this may not be possible for some CR structures (M, θ, J) with $L_\theta > 0$. What is required is a positive mass theorem which we describe.

Given (M, θ, J) with $L_\theta > 0$, the Green's function for $L_\theta G(p, \cdot) = \delta_p$ exists with pole at $p \in M$. To describe the asymptotic behaviour of $G(p, \cdot)$ near the pole, Jerison and Lee [25] showed that after making suitable conformal change of contact form, there will exists local coordinate system (x, y, t) with p at the origin so that denoting by

$$\mathring{\theta} = dt + izd\bar{a} - i\bar{z}dz$$

$$\mathring{\theta}^1 = \sqrt{2}dz, \qquad \mathring{\theta}^{\bar{1}} = \sqrt{2}d\bar{z}$$

the new contact form $\hat{\theta}$ and local frame $\hat{\theta}^1$ are of the form:

$$\hat{\theta} = (1 + O(\rho^4))\mathring{\theta} + O(\rho^5)dz + O(\rho^5)d\bar{z}$$

$$\hat{\theta}^1 = \sqrt{2}(1 + O(\rho^4))dz + O(\rho^4)d\bar{z} + O(\rho^3)\mathring{\theta}$$

and the Green's function satisfies

$$G_p = \frac{1}{2\pi\rho^2} + A + O(\rho) \text{ near } p.$$

Consider the "blowup" of $M\backslash\{p\}$ with

$$\theta = G_\rho^2 \hat{\theta}.$$

Making use of the inversion

$$z_* = \frac{z}{v}, \quad t_* = \frac{t}{|v|^2}, \quad v = t + i|z|^2, \quad \rho_* = \frac{1}{\rho},$$

we find

$$\theta = \left(\frac{1}{(2\pi)^2} + \frac{1}{\pi} A\rho_*^{-2} + O(\rho_*^{-3}) \right) (\mathring{\theta})_* + O(\rho_*^{-3}) dz_*$$

$$+ O(\rho_*^{-3}) d\bar{z}_* \tag{AF_1}$$

$$\theta^1 = \left(\sqrt{2}\frac{Az_*}{v_*^2} + O(\rho_*^{-5}) \right) (\mathring{\theta})_* + O(\rho_*^{-*}) d\bar{z}_*$$

$$+ \left(-\frac{1}{\sqrt{2\pi}} \cdot \rho_*^2 \frac{\bar{v}_*}{v_*^2} - A\sqrt{2}\frac{\bar{v}_*}{v_*^2} + O(\rho_*^{-3}) \right) dz_*.$$

Rotate θ^1 by $e^{i\psi} := -\frac{v_*^3}{\rho_*^6}$ we find

$$\theta_n^1 = \left(-\sqrt{2}\frac{Az_* v_*}{\rho_*^6} + O(\rho_*^{-5}) \right) (\mathring{\theta})_* + O(\rho^{-4}) d\bar{z}_*$$

$$+ \left(\frac{1}{\sqrt{2\pi}} + A\sqrt{2}\rho_*^{-2} + O(\rho_*^{-3}) \right) dz_* \tag{AF_2}$$

This motivates the definition of an asymptotically Heisenberg end i.e. (N^3, J, θ) is called asymptotically flat if $N = N_0 \cup N_\infty$ where N_0 compact and N_∞ is diffeomorphic to $H|'\backslash B_{\rho_0}$ in which (J, θ) is close to $(J_0, \mathring{\theta})$ in this sense i.e. AF_1 and AF_2 hold.

Define the mass for AF manifold N to be

$$m(J, \theta) \overset{\triangle}{=} \lim_{\wedge \to \infty} i \oint_{S_\wedge} \omega_{\bar{1}}^1 \wedge \theta$$

where $S_\wedge = \{\rho = \wedge\}$.

Remarks

1. It follows that, for a family $J^{(s)}$ of CR-structures which are AF with fixed θ,

$$\frac{d}{ds}|_{s=0} \left(-\int_N R(s)\theta \wedge d\theta + m(J(S), \theta) \right) = \int_N (A_{11}E_{\bar{1}\bar{1}} + A_{\bar{1}\bar{1}}E_{11})\theta \wedge d\theta$$

where

$$J = 2E = 2E_{11}\theta^1 \otimes Z_{\bar{1}} + 2E_{\bar{1}\bar{1}}\theta^{\bar{1}} \otimes Z_1.$$

2. If (N, θ, J) arises out of a compact (M, θ, J) as we did then

$$m(J, \theta) = 48\pi^2 A.$$

Theorem ([14]) *Let (N, θ, J) be an asymptotically flat CR manifold, suppose $L_\theta > 0$, and $P \geq 0$ then the mass $m \geq 0$ and equality can hold only if (N, θ, J) is biholomorphic to the Heisenberg group $\mathbb{H}^1$.*

Idea of Proof Look for a function $\beta : N \to \mathbb{C}$ smooth such that

$$\beta = \bar{z} + \beta_{-1} + O(\rho^{-2+\epsilon}) \text{ near } \infty$$

and

$$\Box_b \beta = O(\rho^{-4}),$$

where β_{-1} is a term of homogeneity ρ^{-1} s.t.

$$(\beta_{-1})_{,\bar{1}} = -2\sqrt{2}\frac{\pi A}{\rho^2} - \frac{\sqrt{2}A}{|z|^2 + it} \tag{*}$$

then

$$\frac{2}{3}m(J, \theta) = \int_N \{-|\Box_b \beta|^2 + z|\beta_{,\bar{1}\bar{1}}|^2 + 2R|\beta_{1\bar{1}}|^2 + \frac{1}{2}P\beta \cdot \bar{\beta}\}\theta \wedge d\theta.$$

We find β_{-1} by "hand": choose a cut off function $\eta : \mathbb{H}^1 \to \mathbb{R}$

$$\eta(z, t) = 0 \text{ in a neighborhood of } (0, 0) \tag{4.6.1}$$

$$\eta(z, t) = 1 \text{ in a neighborhood of } \infty \tag{4.6.2}$$

to solve for $\overset{\circ}{\Box}_b \beta_{-1} = -\eta f$ near ∞ where

$$f = 4\pi A\frac{\bar{z}(|z|^2 + it)}{\rho^6} = \Box_b \bar{z} + O(\rho^{-4})$$

making use of solvability of $\overset{\circ}{\Box}_b$ on $\mathbb{H}^1$:

$$\overset{\circ}{\Box}_b K = K\overset{\circ}{\Box}_b = I - S \text{ on } L^2(\mathbb{H}^1)$$

where

$$Kh = -h * \Phi, \; \Phi = \frac{1}{8\pi^2} \log(\frac{|z|^2 - it}{|z|^2 + it}) \cdot (|z|^2 - it)^{-1}.$$

Then it follows that

$$\beta_{-1} = K(-\eta f) + O(\rho^{-2+\epsilon}).$$

and (*) holds.

Existence of a Solution $\Box_b \beta = 0$ This follows from the work of Hsiao and Yung [24] where they studied a weighted $\Box_b$ problem

$$\Box_{b,1} := G_\rho^2 \Box_b, \; m_1 := G_\rho^{-2} \theta \wedge d\theta$$

$$\Box_{b,1} : \text{Dom} (\Box_{b,1}) \subseteq L^2(m_1) \to L^2(m_1) \text{ has closed range}$$

and

$$\Box_{b,1} K + S = I \text{ on } L^2(m_1)$$

for each ϵ in $(0, 2)$ such that

$$\Box_{b,1} K + S = Id \text{ on } \mathcal{E}(\rho^{-2+\epsilon})$$

$$K : \mathcal{E}(\rho^{-2+\epsilon}) \to \mathcal{E}(\rho^\epsilon) \; S : \mathcal{E}(\rho^{-2+\epsilon}) \to \mathcal{E}(\rho^{-2+\epsilon})$$

where $\mathcal{E}(\rho^\mu)$ consists of functions u satisfying $|Z^{(\alpha)} u| \leq c_{\alpha,\mu} \rho^{\mu - |\alpha|}$.
This shows $m(J, \theta) \geq 0$.
When $m(J, \theta) = 0$ the mass formula shows

$$\beta_{\bar{1}\bar{1}} = 0, \beta_{,\bar{1}1} = 0, \; P(\beta) = 0$$

We conclude $R \equiv 0$ but still need to show $A_{11} = 0$.
Let φ_s be the flow generated by the Reeb vector field T, and let

$$J(s) = \varphi_s^* J \text{ i.e. } \dot{J} = 2A_{J,\theta}.$$

Since this deformation is pull back by φ_s, its underlying CR structure is biholomorphic to the original (J, θ), hence the condition $P_{(s)} \geq 0$ is preserved.

Making use of $A_{11,\bar{1}\bar{1}} = O(s\rho^{-8})$ for s small and ρ large we find

$$\frac{d}{ds} R_{J_s,\theta} = -2|A_{11}|^2 + i(A_{11,\bar{1}\bar{1}} - A_{\bar{1}\bar{1},11})$$

$$\frac{d}{ds}(A_{11})_{J_{(s)},\theta} = -i A_{11,0}.$$

We see that

$$R_{J_{(s)},\theta} \geq -cs \text{ on } N$$

$$\geq -c\frac{s}{\rho^8} \text{ near } \infty.$$

Therefore we can solve v_s, decaying to zero near ∞

$$-4\triangle_b v_s + R_{J_{(s)},\theta} v_s = R_{J_{(s)},\theta} \text{ on } N$$

let $u_s = 1 - v_s$, then $(N, J_{(s)}, u_s^2\theta)$ is scalar flat and asymptotic flat.

$$\Rightarrow u_s = 1 - \frac{1}{32\pi\rho^2} \int_N R_{J_{(s)},\theta} u_s\theta \wedge d\theta + O(\rho^{-3})$$

$$\Rightarrow m_s = -\frac{3}{4} \int_N R_{J_{(s)},\theta} u_s\theta \wedge d\theta.$$

Differentiating in s:

$$\frac{d}{ds}\Big|_{s=0} m_s = \frac{3}{2} \int |A_{11}|^2\theta \wedge d\theta > 0 \text{ unless } A_{11} \equiv 0.$$

This implies for $s < 0$ and small $m_s < 0$. This is a contradiction to the first part of this theorem.

Remarks

1. It is possible to show that $\exists$ small perturbation $J_{(s)}$ of the standard CR structure on S^3 which has negative mass. Indeed, in a recent preprint [15], it is shown that the Rossi sphere has negative mass and that the Sobolev quotient is never attained.
2. We believe that the Rossi sphere with $Z_1^{(t)} = Z_1 + t\bar{Z}_1$ for t small have negative mass. In fact in on-going work with Cheng and Malchiodi that for t small, every minimizing sequence for the Sobolev quotient $\frac{\int L_\epsilon u \cdot u}{\|u\|_4^2}$ must blow up. So that the infimum for the Sobolev quotient is never attained.
3. It is quite likely this is true for a large class of perturbations of J.

4. In dimensions $2n + 1 \leq 7$, and locally spherical CR structures. Cheng, Chiu and I showed [11] the developing map is injective, and the CR mass is positive along the same lines as the Schoen–Yau argument.

4.7   The Q-Prime Curvature Equation

In Sect. 4.4 we introduced the P-prime operator and Q-prime curvature equation. For a pseudo-Einstein manifold (M^3, J, θ) the Q' curvature is given by

$$Q' := -2\Delta_b R + R^2 - 4|A_1|^2,$$

under conformal change of contact form

$$\tilde{\theta} = e^\sigma \theta,$$

where σ is pluriharmonic, we have

$$e^{2\sigma} \tilde{Q}' = Q' + P'\sigma + \frac{1}{2} P(\sigma^2).$$

In analogy with Gursky's result [22] about total Q-curvature on a Riemannian 4-manifold we have the following:

Theorem ([8]) *Let $(M^3, \bar{\theta}, J)$ be a pseudo-Einstein manifold with positive CR Yamabe constant and non-negative Paneitz operator. Given any $p \in M$, it holds that*

$$\int_M Q' = 16\pi^2 - 4\int_M G_L^4 |A_{11}|_{\bar{\theta}}^2 - 12\int \log(G_L) P_4 \log(G_L)$$

where G_L is the Green's function for the CR conformal Laplacian with pole at p and $\tilde{\theta} = G_L^2 \theta$. In particular,

$$\int_M Q' \leq 16\pi^2$$

with equality if and only if (M^3, J) is CR equivalent to the standard 3-sphere.

We indicate two proofs of this result, the first one depends on the positive mass theorem and an identity relating the change of the Q'-curvature under conformal change of contact form $\tilde{\theta} = e^\sigma \theta$ where σ is not necessarily pluriharmonic. While the second argument is elementary and hence more transparent.

If the background contact form θ is pseudo-Einstein, and $\tilde{\theta} = e^\sigma \theta$ is the minimizer of the Sobolev quotient so that normalizing the values of $\tilde{\theta}$ to be that of the standard S^3, then

$$R(\tilde{\theta}) = \text{constant} \leq 2.$$

Hence the formal expression

$$\int R^2(\tilde{\theta}) - 4|A_{11}(\tilde{\theta})|^2 \hat{\theta} d\tilde{\theta} \leq \int R^2(\tilde{\theta})\hat{\theta} d\tilde{\theta} \leq \int R^2(\text{standard sphere})dV = 16\pi^2.$$

On the other hand in [7] we found

$$\int Q'(\tilde{\theta})\hat{\theta} \wedge d\tilde{\theta} = \int Q'(\theta)\theta \wedge d\theta + 3 \int \sigma P\sigma\theta \wedge d\theta.$$

Therefore we have

$$\int Q'(\theta)\theta \wedge d\theta \leq 16\pi^2,$$

with equality holding iff the Yamabe invariant of (M^3, J) agrees with the standard sphere, hence it is biholomorphic to the standard sphere.

The second argument works for boundaries of strictly pseudoconvex domains in $\mathbb{C}^2$ which have a contact form θ defined $\theta = Im\bar{\partial}u$ where u is an approximate solution of Fefferman's equation:

$$J[u] = \det \begin{bmatrix} u & \frac{\partial u}{\partial \bar{z}_j} \\ \frac{\partial u}{\partial z_i} & \frac{\partial^2 u}{\partial z_i \partial \bar{z}_j} \end{bmatrix} = 1 + O(u^3).$$

We will consider the contact form $\tilde{\theta} = G_p^2\theta$ where G_p is the Green's function for L_θ with pole at p. Instead of conformal normal coordinates, we work with Moser's coordinates [19]. That is, there is a local biholomorphic change of coordinate near p, say (z, w) with p at the origin so that $\partial\Omega$ is given by, writing $w = u + iv$,,

$$v = |z|^2 + E(u, z, \bar{z}),$$

where

$$E(u, z, \bar{z}) = +c_{42}(u)z^4\bar{z}^2 + c_{24}(u)z^2\bar{z}^4 + c_{33}(u)z^3\bar{z}^3 + O(\rho^7).$$

Let r denote

$$\frac{1}{2i}(w - \bar{w}) - |z|^2 - E(u, z, \bar{z}):$$

then

$$J[r] = 1 + O(\rho^4).$$

Lee-Melrose's asymptotic expansion [27] reads

$$u \sim r \sum_{k \geq 0} \eta_k (r^3 \log r)^k \text{ near } \partial\Omega = \{r = 0\},$$

with

$$\eta_k \in C^\infty(\bar{\Omega})..$$

Thus for N large, $u \sim r \sum_{n=0}^{N} \eta_k (r^3 \log r)^k$ has many continuous derivatives in Ω and vanish to high order at $\partial\Omega$. Hence

$$J[r\eta_0] = 1 + O(\rho^4)$$

$$\eta_0 = 1 + O(\rho^4)$$

$$u \sim r\eta_0 + \eta_1 r^4 \log r + \text{h.o.t}$$

$$\sim r + O(\rho^6).$$

We compute

$$\left.\begin{array}{l} \theta \ \ = (1 + O(\rho^4))\mathring{\theta} + O(\rho^5)dz + O(\rho^5)d\bar{z} \\ \theta^1 = O(\rho^3)\mathring{\theta} + (1 + O(\rho^8))dz + O(\rho^8)d\bar{z} \\ Z_1 = \mathring{Z}_1 + O(\rho^5)\partial_u \\ \omega_1^1 = O(\rho^2)\mathring{\theta} + O(\rho^3)dz + O(\rho^7)d\bar{z} \\ A_1^1 = O(\rho^2), \ R = O(\rho^2) \\ h_{1\bar{1}} = 1 + O(\rho^6), \ h^{1\bar{1}} = 1 + O(\rho^6). \end{array}\right] \qquad (4.7.1)$$

Claim 1 $P'(\log G_L) = 8\pi^2 S_p + $ a bounded function, where $S_p = S(p, \cdot)$ is the kernel of the orthogonal projection $\pi : L^2 \to L^2 \cap \mathcal{P}$ onto pluriharmonics.

Making use of a similar expansion for $\triangle_b$, write

$$G_L = \frac{1}{2\pi\rho^2} + \omega,$$

we find that $L\omega$ is bounded function near p.

It then follows that

$$\log(G_L) = \log(\frac{1}{2\pi\rho^2} + \omega)$$

$$= \log\frac{1}{2\pi\rho^2} + \log(1 + 2\pi\rho^2\omega)$$

$$P'\log(G_L) = 8\pi^2 S_p + \left\{4(\Delta_b^2 - \mathring{\Delta}_b^2) - 8Im(\nabla^1(A_1^{\bar{1}}\nabla_1))\right.$$

$$\left. -4Re(\nabla^1(R\nabla_1))\right\}\left(\log\frac{1}{2\pi\rho^2}\right) + P'\log(1 + 2\pi\rho^2\omega).$$

The smoothness of ω then implies that the third term is bounded.

The second term is also bounded by (4.7.1). This proves Claim 1.

Claim 2 $P((\log G_L)^2) = 8\pi^2(\delta_p - S_p) +$ a bounded function. This follows from:

$$\mathring{P}_3(\log\rho) = 0$$

because ρ is the absolute value of a holomorphic function. Hence

$$P_3(\log G_L) = O + (P_3 - \mathring{P}_3)(\log\frac{1}{2\pi\rho^2}) + P_3\left(\log(1 + 2\pi\rho^2\omega)\right) = O(\rho).$$

In addition,

Claim 3 $(\log G_L)P(\log G_L)$ blows up like $\log\rho$ near p.

Therefore, it is integrable.

The transformation rule shows that away from the pole p:

$$-4G_L^4|\hat{A}_{11}|^2 = Q' + 2P'(\log G_L) + 2P((\log G_L)^2) - 4(\log G_L)P(\log G_L)$$

$$- 64Re(J^1 \log G_L)(P_3(\log G_L))_1,$$

so we find

$$G_L^4|\hat{A}_{11}|^2 = O(\rho^4).$$

Thus in the distribution sense

$$2P'(\log G_L) + 2P((\log G_L)^2) = 16\pi^2\delta_p - Q' - 4G_L^4|\tilde{A}_{11}|^2 + 4(\log G_L)P(\log G_L)$$

$$+ 64Re(\nabla^1 \log G_L)(P_3(\log G_L))_1.$$

Apply this equation to the constant function 1 gives

$$\int Q' = 16\pi^2 - 4 \int G_L^4 |\tilde{A}_{11}|^2 - 12 \int (\log G_L) P(\log G_L) \leq 16\pi^2.$$

Observe finally equality holds iff $\tilde{A}_{11} \equiv 0$ and $\log G_L$ is pluriharmonic. Since $\tilde{R} \equiv 0$, the space $(M \setminus p, \tilde{\theta})$ is CR flat, and is simply connected at ∞, hence is globally isometric to the Heisenberg group.

4.7.1 An Existence Result for the Q'-Curvature Equation

There is a variational functional whose critical point when restricted to the pluriharmonics is the Q'-curvature equation, modulo a Lagrange multiplier. Currently we do not know how to determine the Lagrange multiplier. To get around this difficulty we introduce the modified P-prime operator $\bar{P}' := \tau P' : \mathcal{P} \to \mathcal{P}$ where τ is the L^2 projection to $\mathcal{P}$. Consider the functional $II : \mathcal{P} \to \mathbb{R}$

$$II(\sigma) = \int \sigma \bar{P}' \sigma + 2 \int \bar{Q}' \sigma - (\int \bar{Q}') \log \int e^{2\sigma}.$$

Theorem ([5]) *Let (M^3, θ, J) be a compact pseudo-Einstein 3-manifold with $L_\theta > 0$ and $P_\theta \geq 0$. Suppose*

$$\int \bar{Q}' \theta \wedge d\theta < 16\pi^2,$$

then there exists a function $\sigma \in \mathcal{P}$ minimizing the functional II. Moreover, the contact form $\tilde{\theta} = e^\sigma \theta$ has $\bar{Q}'$ equal to a constant.

Remark The conclusion cannot be strengthened to $Q' = $ constant, in fact there are contact forms on $S^1 \times S^2$ with its spherical CR structure such that will have $\bar{Q}' = 0$ but Q' is non-constant.

The main analysis is a Moser–Trudinger inequality which is deduced from the asymptotics for the Green's function of the pseudo-differential operator $(\bar{P}')^{\frac{1}{2}}$. Fix a point $p \in M$. Let (z, t) be the CR normal coordinates defined in a neighborhood of $p = (0, 0)$.

Let $E(\rho^k)$ denote the class of functions $g \in C^\infty(M \setminus \{p\})$ satisfying

$$|\partial_z^p \partial_{\bar{z}}^q \partial_t^r g(z, t)| \leq C\rho(z, t)^{k-p-q-2r} \text{ near } 0.$$

The bulk of the work is to establish the following:

Claim There exists a $B_p \in C^\infty(M\setminus\{p\})$ s.t.

$$B_p - \frac{1}{\rho^2} \in E(\rho^{-1-\epsilon})$$

for all $0 < \epsilon < 1$ and

$$G_p = \tau B_p \tau.$$

Idea of Proof One has

$$(\bar{P}')^{-\frac{1}{2}} = c \int_0^\infty t^{-\frac{1}{2}} (\bar{P}' + t + \pi)^{-1} dt$$

from spectral theory and

$$G_p = \bar{P}^{-\frac{1}{2}} \tau \delta_p \tau.$$

The analysis involved a detailed study of the family G_t of PDO of order -2 depending continuously in t s.t.

$$(E_2 + t)G_t = I + F_t$$

where E_2 satisfies $\bar{P}' = \tau E_2$, E_2 being a classical PDO of order 2 and F_t a smoothing operator depending continuously on t.

Proof of Theorem Making use of the claim, we established the following Moser–Trudinger inequality

$$\log \int e^{2(\sigma - \bar{\sigma})} \leq c + \frac{1}{16\pi^2} \int \sigma \bar{P}' \sigma.$$

This in turn yields estimate for minimizing sequence of the functional II.

4.8   An Isoperimetric Inequality

An interesting application of the Q-prime curvature integral is the following [32]:

Theorem *On $\mathbb{H}^1$, let σ be a pluriharmonic function on $\mathbb{H}^1$, such that $e^\sigma \theta$ is a complete pseudo-Einstein contact form, suppose that $\lim_{p \to \infty} R(e^\sigma \theta) \geq 0$ and*

$Q'(e^\sigma \theta) \geq 0$ and $\int Q'\theta \wedge d\theta < C_1$. *Then for any bounded domain* $\Omega \subset \mathbb{H}^1$

$$vol\,(\Omega) \leq C\,area\,(\partial\Omega)^{\frac{4}{3}}$$

where the constant C *depends only on the difference* $C_1 - \int Q'\theta \wedge d\theta$.

Remarks

1. This is a weaker version of an analogous result of Wang [31] about Q-curvature integral on $\mathbb{R}^n$, where the Q-curvature is given by

$$(\triangle)^{\frac{n}{2}} u = Q e^{1-u}$$

   for the conformal metric $g = e^{2u}|dx|^2$.
2. The constant C_1 is critical in that there exists example where $\int_{\mathbb{H}'} Q'\theta \wedge d\theta = C_1$ and no isoperimetric constant exists.
3. This is a result in harmonic analysis where the assumptions imply that $\omega = e^{2\sigma}$ is an A_1-weight, i.e. as a measure

$$\frac{\omega(B)}{|B|} \lesssim \omega(x), \qquad x \text{ being the center of the ball } B.$$

Idea of Proof The first step is to show that $\tilde{\theta} = e^\sigma \theta$ is a normal contact form i.e.

$$\sigma(x) = \frac{1}{C_1} \int_{\mathbb{H}'} \frac{\log \rho(y)}{\rho(y^{-1} \cdot x)} Q'(y) e^{2\sigma(y)} dv(y) + C.$$

The rest of the argument is to show that $\omega = e^{2\sigma}$ is an A_1-weight.

A Counterexample The contact form $\hat{\theta} = \frac{1}{\rho^2}\theta$ on $\mathbb{H}^1 \setminus \{0\}$ is clearly not an A_1 weight and it does not satisfy an isoperimetric inequality. It is easy to construct a smoothing of $\tilde{\theta}$ so that it satisfies the assumptions of the theorem. This shows that the constant C_1 is sharp.

4.9 Geometry of Surfaces in the Heisenberg Group

In this last topic we discuss the local invariants of a surface Σ^2 in the Heisenberg group $\mathbb{H}^1$, the p-mean curvature equation and several global questions of interest to analysts.

In the following I will summarize briefly the local invariants of a surface in the Heisenberg group $\mathbb{H}^1$. Given a smooth surface $\Sigma \subset \mathbb{H}^1$ at a generic point p the tangent space and the contact plane are distinct. We let e_1 be a unit vector whose span $\langle e_1 \rangle = T_p\Sigma \cap \gamma_1$ and p will be called a regular point. Nearby point q will also be regular and the e_1 vector field exists locally. It determines integral curves γ, $(\dot{\gamma} = e_1)$ called characteristic curves. The equation $\nabla_{\dot{\gamma}} \dot{\gamma} = HJ\dot{\gamma}$ defines the p-mean curvature H for the surface. Along the regular part of the surface we have the local framing $\{e_1, e_2 = Je_1, T\}$. Since T and e_2 are both transverse to $T_p\Sigma$, there exist unique α so that

$$T + \alpha e_2 \in T_p\Sigma.$$

α is called the angle function. The associated dual frame $\{e^1, e^2, \theta\}$ exists in a neighborhood of p. It is natural to define the area element as $|e^1 \wedge \theta|$, in fact its integral gives the 3-dimensional Hausdorff measure of Σ.

The p-mean curvature equation for a graph $t = u(x, y)$ read as

$$\frac{(u_y + x)^2 u_{xx} - 2(u_y + x)(u_x - y)u_{xy} + (u_x - y)^2 u_{yy}}{((u_x - y)^2 + (u_y + x)^2)^{\frac{3}{2}}} = H.$$

It is a degenerate hyperbolic equation.

Notice that the p-mean curvature is not defined at a characteristic point p on the surface, i.e. where $T_p\Sigma = \xi_p$ or in the case of a graph

$$u_y + x = 0 \text{ and } u_x - y = 0.$$

We summarize the key facts below [12].

When Σ is C^2-smooth and the p-mean curvature satisfy the condition near a singular point $p : |H(q)| = O(\frac{1}{|p-q|})$ then there is the classification:

1. either p is an isolated singular point and the e_1-line field has an isolated singularity at p, and its topological degree is 1; or
2. p is part of a C^1-curve γ consisting of the singular points near p and the characteristic line field e_1 pass transversally through the singular curve; and as a line field, it is not singular at the curve γ.

If Σ^2 is a closed surface in $\mathbb{H}^1$ with bounded p-mean curvature then $\chi(\Sigma^2) \geq 0$. This is the direct consequence of the Hopf index theorem.

Pansu's sphere is obtained by rotating the following curve around the z-axis:

$$X(s) = \frac{1}{2\lambda} \sin(2\lambda s)$$

$$Y(s) = \frac{1}{2\lambda}(-1 + \cos(2\lambda s))$$

$$Z(s) = \frac{1}{2\lambda}(s - \frac{1}{2\lambda} \sin(2\lambda s))$$

Theorem ([28]) *If Σ^2 is smooth C^2 solution of the p-mean curvature equation with $H = constant > 0$, then it is congruent to the Pansu sphere. It has two singular points $(0, 0, 0)$ and $(0, 0, \pi)$.*

Elliptic Approximation One way to deal with the lack of ellipticity in this problem is to approximate the Heisenberg geometry by a family of Riemannian metrics, and then to understand the limiting behavior of the solutions of the Riemannian mean curvature equation. This is difficult to do because the Heisenberg is not the metric limit of Riemannian spaces with Ricci bounded from below.

Conditions for a Weak Solution

$$
\begin{cases}
F[u] = \int_\Omega |\nabla u + F| + Hu\,dx \text{ in } \Omega \subset \mathbb{R}^2 \\[2mm]
\text{where} \qquad F = \begin{bmatrix} -y \\ x \end{bmatrix} \\[4mm]
\text{Euler equation} \qquad div \frac{\nabla u + F}{|\nabla u + F|} = H
\end{cases}
\tag{II}
$$

We say $u \in W^1(\Omega)$ is a weak solution of the p-mean curvature equation if

$$
\int_{S[u]} |\nabla \varphi| + \int_{\Omega \setminus S[u]} N(u) \cdot \nabla \varphi + \int H\varphi \geq 0, \text{ for all test functions } \phi.
$$

One surprising feature of this equation is the following equivalence:

Theorem *$u \in W^{1,1}(\Omega)$ is a weak solution if and only if u is a minimizing solution.*

Analogue of the Codazzi Equation Let V denote the vector field $T + \alpha e_2$: then [13]

$$
e_1 e_1(\alpha) = -6\alpha e_1(\alpha) + V(H) - \alpha H^2 - 4\alpha^3
$$

along each characteristic curve.

It is the analogue of the Codazzi equation because it is part of a system of equations that characterize a smooth surface in the Heisenberg group as an abstract surface with a smooth foliation of curves just like the Gauss and Codazzi equation characterized a smooth surface in $\mathbb{R}^3$

Invariant Surface Area Functionals
In a pseudo-Hermitian manifold (M^3, θ, J) let Σ^2 be a smooth surface without singular points. The following two area functionals are conformally invariant:

$$
dA_1 = |e_1(\alpha) + \frac{1}{2}\alpha^2 - ImA_{11} + \frac{1}{4}R + \frac{1}{6}H^2|^{\frac{3}{2}}\theta \wedge e^1
$$

$$
dA_2 = \left\{ V(\alpha) + \frac{2}{3}\left[e_1(\alpha) + \frac{1}{2}\alpha^2 - ImA_{11} + \frac{1}{4}R \right] H + \frac{2}{27}H^3 \right.
$$

$$
\left. + Im\left(\frac{1}{6}R_{\bar{1}} + \frac{2}{3}iA_{\bar{1}\bar{1},1} \right) + \alpha ReA_{\bar{1}\bar{1}} \right\} \theta \wedge e^1
$$

They remain pointwise conformally invariant under conformal change of contact form $\hat{\theta} = e^{2u}\theta$.

These invariants were first given by J. Cheng in terms of the Chern connection, and more recently by Cheng et al. [16] in terms of the local invariants developed more recently.

Examples in the Heisenberg Group

1. The vertical plane $y = 0$ it is easy to see that $H = 0$ and $\alpha = 0$, hence it is a minimizer for the energy integral $\int_{\Sigma} dA_1 = E_1[\Sigma]$, it is also critical for E_2.

2. The shifted sphere: $(|z|^2 + \frac{\sqrt{3}}{2}\rho_0^2)^2 + 4t^2 = \rho_0^4$ (for any positive ρ_0). It is an easy computation to show $dA_1 \equiv 0$ hence it is a closed example having the topology of the 2-sphere. It is an open question whether this is the unique minimizer (up to a biholomorphic transformation of the Heisenberg group).

3.

$$t = \frac{\sqrt{3}}{2}|z|^2 \qquad \text{the paraboloid}$$

Again

$$e_1(\alpha) + \frac{1}{2}\alpha^2 + \frac{1}{6}H^2 = 0$$

4. The Clifford Torus

$$\rho_1 = \rho_2 = \frac{\sqrt{2}}{2} \subset S^3$$

where

$$z_1 = \rho_1 e^{i\varphi_1}, z_2 = \rho_2 e^{i\varphi_1}.$$

This surface is p-minimal in S^3, i.e. $H = 0$. It is, after a bit of computation, a critical point for E. Again it is an open question whether this is the unique minimizer for E_1 under the topological constraint to be a torus.

Connections to the Singular Yamabe Problem

Consider change of contact form $\hat{\theta} = u^{-2}\theta$, we consider $\Omega \subset \mathbb{H}'$ the singular CR Yamabe problem:

$$\begin{cases} -\Delta_b(\frac{1}{u}) = -(\frac{1}{u})^3 & \text{on } \Omega \\ u = 0 & \text{on } \partial\Omega = \Sigma \end{cases} \qquad (4.9.1)$$

It is relatively easy to see the existence of solutions to this equation by means of the strong maximum principle.

The interesting question is when is the solution smooth at the boundary.

We develop the solution u in Taylor series expansion in powers of ρ, a defining function for $\partial\Omega$.

Thus

$$u(x, p) = c(x)\rho + v(x)\rho^2 + w(x)\rho^3 + z(x)\rho^4 + l(x)\rho^5 \log \rho + h(x)\rho^5 + O(\rho^6)$$

Similar to the expansion for the Riemannian singular Yamabe solution expansion of R. Graham we are able to determine, after an extended calculation, these coefficients c, v, w, z, l.

It turns out that l is related to the volume renormalization coefficient L and the functional E_2.

Consider the volume expansion

$$\text{Vol}\{\rho > \epsilon\} = \int_{\{\rho>\epsilon\}} u^{-4} \langle e_2, v\rangle d\mu_{\Sigma_\rho} d\rho.$$

$$\text{Vol}(\{\rho > \epsilon\}) = c_0\epsilon^{-3} + c_1\epsilon^{-2} + c_2\epsilon^{-1} + L \log \frac{1}{\epsilon} + V_0 + o(1)$$

with

$$c_0 = \frac{1}{3} \int_\Sigma \theta \wedge \theta^1$$

$$c_1 = -\frac{1}{6} \int_\Sigma H\theta \wedge \theta^1$$

$$c_2 = \frac{1}{3} \int \left(5e_1(\alpha) + 10\alpha^2 + \frac{1}{6}H^2 - a_2\right)\theta \wedge \theta^1$$

$$L = \int_\Sigma \underbrace{\frac{1}{6}e_1e_1(H) + 4V(\alpha) + \alpha e_1(H) + 2He_1(\alpha) + \frac{4}{27}H^3}_{V^{(3)}}$$

and where

$$[T, e_1] = -a_1e_1 + [a_2 + \omega(T)]e_2$$

$$[e_2, T] = -a_1e_2 + (-a_2 + \omega(T))e_1$$

It turns out $\frac{1}{2}V^{(3)}$ differs from dA_2 by an exact form

$$dA_2 - \frac{1}{2}V^{(3)} = \frac{1}{12}d(e_1(H)\theta) - d(\alpha e^1) + \frac{1}{3}d(\alpha H\theta) + \frac{1}{2}d(a_1\theta)$$

and l is the Euler equation in L.

Therefore we find

Theorem ([16]) *The solution u to the equation* $(\delta, 1)$ *is smooth up to order 5 if and only if* $\partial\Omega = \Sigma$ *is a critical point for the energy functional* E_2.

It is an interesting question to ask where does the energy functional E_1 come from.

Example The upper Heisenberg $\tilde{\theta} = \frac{1}{y^2}\theta$. In this case, the function $u = y$ is smooth at the boundary.

Euler Equations of E_1 and E_2

$$\delta E_1 = l = He_1e_1H + 3e_1V(H) + e_1(H)^2 + \frac{1}{3}H^4 + 3e_1(\alpha)^2 + 12\alpha^2e_1(\alpha) + 12\alpha^4$$

$$- \alpha He_1(H) + 2H^2e_1(\alpha) + 5\alpha^2H^2 + \frac{3}{2}R(e_1(\alpha)) + \frac{2}{3}H^2 + 5\alpha^2 + \frac{1}{2}R.$$

$$\delta E_2 = |H_{cr}|^{-\frac{1}{2}}\left\{-\frac{1}{4}sqn(H_{cr})fe_1(H_{cr}) + \frac{1}{2}e_1(|H_{cr}|f) + \frac{3}{2}(H_{cr})f\alpha\right.$$

$$\left. +H_{cr}\left[\frac{9}{2}V(\alpha) + 3HH_{cr} - \frac{1}{6}H^3\right]\right\}.$$

where

$$H_{cr} = e_1(\alpha) + \frac{1}{2}\alpha^2 + \frac{1}{6}H^2$$

$$f = \frac{\frac{1}{2}e_1(H)H_{cr} + \frac{3}{4}V(H_{cr} + \frac{1}{4}He_1(H_{cr}) - \frac{1}{2}\alpha HH_{cr}}{2|H_{cr}|}.$$

Open Questions In these lectures we dealt with the extremals of the following two Sobolev inequalities:

$$c_1||u||_{\frac{4}{3}} \le \int |\nabla_b u|$$

$$c_2||u||_4 \le \int |\nabla_b u|^2$$

It is natural to wonder about

$$c_1||u||_q \le \int |\nabla_b u|^p ?$$

and what are the level sets of critical functions? It appears likely that there are fractional order conformally covariant operators which may shed light on sharp form of inequalities of these type.

As for geometric applications for some of the analysis discussed here we hope to be able to study the following type of questions. A basic question is:

$$0 < \int_{M^3} Q' \theta \wedge d\theta \Rightarrow M^3 = S^3 / \Gamma, \quad \text{where } \Gamma \text{ is a finite subgroup of } U(2),$$

and whether $(S^3 / \Gamma, \theta_{can}, J_{can})$ realizes maximal value of $\int Q'$. for this topology.

Similarly if M^3 is the unit tangent bundle of a hyperbolic surface Σ, it is expected that it also realizes the maximal value of

$$\int Q' \theta \wedge d\theta$$

for the given topology.

References

1. J. Bland, Contact geometry and CR structures on S^3. Acta Math. **172**, 1–49 (1994)
2. L. Boutet de Monvel, Integration des equations Cauchy–Riemann induites formelles, in *Seminaire Goulaouic–Lions-Schwartz 1974–1975* (Centre de Mathematiques, Ecole Polytechnique, Paris, 1975)
3. D.M. Burns, Jr., Global behavior of some tangential Cauchy–Riemann equations, in *Proceedings of the Conference on Partial Differential Equations and Geometry* (Park City, Utah, 1977). Lecture Notes in Pure and Applied Mathematics, vol. 48 (Dekker, New York, 1979), pp. 51–56, No. 9, 14pp
4. D.M. Burns, C. Epstein, A global invariant for three-dimensional CR-manifolds. Invent. Math. **92**(2), 333–348 (1988)
5. S. Case, C.-Y. Hsiao, P. Yang, Extremal metrics for the Q-curvature in three dimensions. C.R. Math. **354**(4), 407–410
6. S. Chanillo, H.-L. Chiu, P. Yang, Embeddability for 3-dimensional Cauchy–Riemann manifolds and CR Yamabe invariants. Duke Math. J. **161**(15), 2909–2921 (2012)
7. J.S. Case, P. Yang, A Paneitz-type operator for CR pluriharmonic functions. Bull. Inst. Math. Acad. Sin. (N.S.) **8**(3), 285–322 (2013)
8. J. Case, J.-H. Cheng, P. Yang, An integral formula for the Q-prime curvature in 3-diemnsional CR geometry. Proc. AMS **147**(4), 1577–1586 (2019)
9. J.S. Case, S. Chanillo, P. Yang, The CR Paneitz operator and the stability of CR pluriharmonic functions. Adv. Math. **287**, 109–122 (2016)
10. S.-C. Chen, M.-C. Shaw, in *Partial Differential Equations in Several Complex Variables*. AMS/IP Studies in Advanced Mathematics, vol. 19
11. J.-H. Cheng, H.-L. Chiu, P. Yang, Uniformazation of spherical CR manifolds. Adv. Math. **255**, 182–216 (2014)
12. J.-H. Cheng, J.-F. Hwang, A. Malchiodi, P. Yang, Minimal surfaces in pseudo-Hermitian geometry. Ann. Sc. Norm. Super. Pisa Cl. Sci. **4**(1), 129–177 (2005)
13. J.-H. Cheng, J.-F. Hwang, A. Malchiodi, P. Yang, A Codazzi-like equation and the singular set for C1 smooth surfaces in the Heisenberg group. J. Reine Angew. Math. **671**, 131–198 (2012)
14. J.-H. Cheng, A. Malchiodi, P. Yang, A positive mass theorem in three dimensional Cauchy–Riemann geometry. Adv. Math. **308**, 276–347 (2017)
15. J.-H. Cheng, A. Malchiodi, P. Yang, On the Sobolev quotient of three dimensional CR manifolds. arXiv:1904.04665

16. J.-H. Cheng, P. Yang, Y. Zhang, Invariant surface area functionals and singular Yamabe problem in 3-dimensional CR geometry. Adv. Math. **335**, 405–465. arXiv:1711.04120
17. J.-H. Cheng, S.T. Yau, On the existence of a complete Kähler metric on noncompact complex manifolds and the regularity of Fefferman's equation. Commun. Pure Appl. Math. **33**(4), 507–544 (1980)
18. J.-H. Cheng, J.M. Lee, The Burns–Epstein invariant and deformation of CR structures. Duke Math. J. **60**(1), 221–254 (1990)
19. S.S. Chern, J.K. Moser, Real hypersurfaces in complex manifolds. Acta Math. **133**, 219–271 (1974)
20. C.L. Fefferman, Monge–Ampère equations, the Bergman kernel, and geometry of pseudoconvex domains. Ann. Math. **103**(2), 395–416 (1976)
21. G.B. Folland, E.M. Stein, Estimates for the $\bar{\partial}_b$ complex and analysis on the Heisenberg group. Commun. Pure Appl. Math. **27**, 429–522 (1974)
22. M.J. Gursky, The principal eigenvalue of a conformally invariant differential operator, with an application to semilinear elliptic PDE. Commun. Math. Phys. **207**(1), 131–143 (1999)
23. K. Hirachi, Invariants of CR structures and the logarithmic singularity of the Bergman kernel, in *Geometric Complex Analysis* (Hayama, 1995) (World Scientific Publishing, River Edge, 1996), pp. 239–247
24. C.-Y. Hsiao, P.-L. Yung, Solving the Kohn Laplacian on asymptotically flat CR manifolds of dimension 3. Adv. Math. **281**, 734–822 (2015)
25. D. Jerison, J.M. Lee, Intrinsic CR normal coordinates and the CR Yamabe problem. J. Differ. Geom. **29**, 303–343 (1989)
26. J. Lee, Pseudo-Einstein structures on CR manifolds. Am. J. Math. **110**, 157–178 (1988)
27. J.M. Lee, R. Melrose, Boundary behavior of the complex Monge–Ampere equation. Acta Math. **148**, 159–192 (1982)
28. M. Ritor, C. Rosales, Area-stationary surfaces in the Heisenberg group H^1. Adv. Math. **219**(2), 633–671 (2008)
29. M. Rumin, Forms diffferentielle sur les varietes de contact. J. Differ. Geom. **39**, 281–330 (1996)
30. Y. Takeuchi, Non-negativity of the CR Paneitz operator for embeddable CR manifolds (2019). arXiv:1908.07672
31. Y. Wang, The isoperimetric inequality and Q-curvature. Adv. Math. **281**, 823–844 (2015)
32. Y. Wang, P. Yang, Isoperimetric inequality on CR-manifolds with nonnegative -curvature. Ann. Sc. Norm. Super. Pisa Cl. Sci. **18**(1), 343–362 (2018)

LECTURE NOTES IN MATHEMATICS   🐎 Springer

Editors in Chief: J.-M. Morel, B. Teissier;

Editorial Policy

1. Lecture Notes aim to report new developments in all areas of mathematics and their applications – quickly, informally and at a high level. Mathematical texts analysing new developments in modelling and numerical simulation are welcome.

   Manuscripts should be reasonably self-contained and rounded off. Thus they may, and often will, present not only results of the author but also related work by other people. They may be based on specialised lecture courses. Furthermore, the manuscripts should provide sufficient motivation, examples and applications. This clearly distinguishes Lecture Notes from journal articles or technical reports which normally are very concise. Articles intended for a journal but too long to be accepted by most journals, usually do not have this "lecture notes" character. For similar reasons it is unusual for doctoral theses to be accepted for the Lecture Notes series, though habilitation theses may be appropriate.

2. Besides monographs, multi-author manuscripts resulting from SUMMER SCHOOLS or similar INTENSIVE COURSES are welcome, provided their objective was held to present an active mathematical topic to an audience at the beginning or intermediate graduate level (a list of participants should be provided).

   The resulting manuscript should not be just a collection of course notes, but should require advance planning and coordination among the main lecturers. The subject matter should dictate the structure of the book. This structure should be motivated and explained in a scientific introduction, and the notation, references, index and formulation of results should be, if possible, unified by the editors. Each contribution should have an abstract and an introduction referring to the other contributions. In other words, more preparatory work must go into a multi-authored volume than simply assembling a disparate collection of papers, communicated at the event.

3. Manuscripts should be submitted either online at www.editorialmanager.com/lnm to Springer's mathematics editorial in Heidelberg, or electronically to one of the series editors. Authors should be aware that incomplete or insufficiently close-to-final manuscripts almost always result in longer refereeing times and nevertheless unclear referees' recommendations, making further refereeing of a final draft necessary. The strict minimum amount of material that will be considered should include a detailed outline describing the planned contents of each chapter, a bibliography and several sample chapters. Parallel submission of a manuscript to another publisher while under consideration for LNM is not acceptable and can lead to rejection.

4. In general, **monographs** will be sent out to at least 2 external referees for evaluation.

   A final decision to publish can be made only on the basis of the complete manuscript, however a refereeing process leading to a preliminary decision can be based on a pre-final or incomplete manuscript.

   Volume Editors of **multi-author works** are expected to arrange for the refereeing, to the usual scientific standards, of the individual contributions. If the resulting reports can be forwarded to the LNM Editorial Board, this is very helpful. If no reports are forwarded or if other questions remain unclear in respect of homogeneity etc, the series editors may wish to consult external referees for an overall evaluation of the volume.

5. Manuscripts should in general be submitted in English. Final manuscripts should contain at least 100 pages of mathematical text and should always include

   - a table of contents;
   - an informative introduction, with adequate motivation and perhaps some historical remarks: it should be accessible to a reader not intimately familiar with the topic treated;
   - a subject index: as a rule this is genuinely helpful for the reader.
   - For evaluation purposes, manuscripts should be submitted as pdf files.

6. Careful preparation of the manuscripts will help keep production time short besides ensuring satisfactory appearance of the finished book in print and online. After acceptance of the manuscript authors will be asked to prepare the final LaTeX source files (see LaTeX templates online: https://www.springer.com/gb/authors-editors/book-authors-editors/manuscriptpreparation/5636) plus the corresponding pdf- or zipped ps-file. The LaTeX source files are essential for producing the full-text online version of the book, see http://link.springer.com/bookseries/304 for the existing online volumes of LNM). The technical production of a Lecture Notes volume takes approximately 12 weeks. Additional instructions, if necessary, are available on request from lnm@springer.com.

7. Authors receive a total of 30 free copies of their volume and free access to their book on SpringerLink, but no royalties. They are entitled to a discount of 33.3 % on the price of Springer books purchased for their personal use, if ordering directly from Springer.

8. Commitment to publish is made by a *Publishing Agreement*; contributing authors of multiauthor books are requested to sign a *Consent to Publish form*. Springer-Verlag registers the copyright for each volume. Authors are free to reuse material contained in their LNM volumes in later publications: a brief written (or e-mail) request for formal permission is sufficient.

Addresses:
Professor Jean-Michel Morel, CMLA, École Normale Supérieure de Cachan, France
E-mail: moreljeanmichel@gmail.com

Professor Bernard Teissier, Equipe Géométrie et Dynamique,
Institut de Mathématiques de Jussieu – Paris Rive Gauche, Paris, France
E-mail: bernard.teissier@imj-prg.fr

Springer: Ute McCrory, Mathematics, Heidelberg, Germany,
E-mail: lnm@springer.com

Printed in the United States
By Bookmasters